Achieving a Leadership Role for Training

How to Use the Baldrige Criteria and ISO Standards to Keep the Training Function Competitive

Judith A. Hale
Odin Westgaard

QUALITY RESOURCES®
A Division of The Kraus Organization Limited
902 Broadway, New York, New York 10010

Most Quality Resources books are available at quantity discounts when purchased in bulk. For more information contact:

Special Sales Department
Quality Resources
A Division of The Kraus Organization Limited
902 Broadway
New York, New York 10010
800-247-8519

Copyright © 1995 Judith A. Hale, Odin Westgaard

All rights reserved. No part of this work covered by the copyrights hereon may be reproduced or used in any form or by any means—graphic, electronic, or mechanical, including photocopying, recording, taping, or information storage and retrieval systems—without written permission of the publisher.

Printed in the United States of America

99   98   97   96   95   10   9   8   7   6   5   4   3   2   1

Quality Resources
A Division of The Kraus Organization Limited
902 Broadway
New York, New York 10010
800-247-8519

The paper used in this publication meets the minimum requirements of American National Standard for Information Sciences—Permanence of Paper for Printed Library Materials, ANSI Z39.48-1984.

ISBN 0-527-76249-0

Library of Congress Cataloging-in-Publication Data

Hale, Judith A.
    Achieving a leadership role for training : how to use the Baldrige criteria and ISO standards to keep training competitive / Judith A. Hale, Odin Westgaard.
      p.   cm.
    Includes index.
    ISBN 0-527-76249-0
    1. Employees—Training of—Quality control.   2. Executives—Training of—Quality control.   3. Total quality management—Study and teaching.
    4. International Organization for Standardization.
  I. Westgaard. Odin.   II. Title.
HF5549.5.T7H283   1995   95-6282
658.3′12404--dc20   CIP

This book is fondly dedicated to Cecil Hale whose thinking continues to inspire us.

Contents

Acknowledgments   **vii**

Introduction   **xi**

Chapter 1   **Training and Quality**   **1**

Chapter 2   **Training and Leadership**   **31**

Chapter 3   **Strategic Planning: Operating Training as a Business**   **53**

Chapter 4   **Optimizing Training Processes**   **83**

Chapter 5   **Training and Standards**   **119**

Chapter 6   **Measurement and Evaluation**   **145**

Index   **173**

Acknowledgments

Quality can't happen without training. Adaptation of Baldrige or ISO standards will only succeed if the training function can support the effort. It can provide new skills, strengthen existing skills, provide consistency across work groups, and ensure cultural integrity. Without a well-managed training function, an organization would be hard-pressed to accomplish any of these tasks.

Training is dependent on the people who provide it. If they are good at what they do, the training will probably accomplish its goals. If they aren't, the results are likely to be inadequate in some way. So, in effect, quality services, quality in production, quality of work life, quality in all of its manifestations depends on the people who provide training.

SUPPORT FROM BUSINESS AND INDUSTRY

This book has many stories about what training managers have done to enhance the quality of their training departments. The stories provide constant reference to the real world. In particular, they speak to the leadership role the training function plays. It tells how training can apply the Baldrige criteria and ISO standards to training itself. I am deeply indebted to colleagues and clients who have so graciously allowed me to talk about them and their efforts. My close friend, Charline Seyfer gave many hours of advice and detailed information about what she and her colleagues are accomplishing at Sandia National Labs. George Pollard, Training Manager at FedEx has been a firm, fair critic and supplier of an important success story about adoption of ISO standards. He has also provided liaison with the International Board of Standards for Training, Performance, and Instruction (IBSTPI). We also want to thank the following people for allowing us to share their work with you: Mark Lindner, Manager of Training, Allstate Commercial Insurance—a division of Allstate Insurance; James Jackson, Manager of Amoco Corporation's Information Technology Training Group; McDonald's Corporation's Purchasing Officers; Anne Marie Laures, Corporate Manager of Training and Development, Walgreens; Don Smith, Manager Corporate Quality, Ford Motor Co.; Janet Gregory and Nellie Hanley, Chairs of IBSTPI's taskforce for certification of instructional designers; and the many others who contributed in one way or another.

These training managers and their organizations deserve much more praise than they have gotten until now. They are the unsung heroes of the quality movement.

MY COLLEAGUES AT HALE ASSOCIATES

I am particularly blessed with coworkers who are not only knowledgeable and experienced, but also willing (anxious even) to help in whatever way I ask. Dave Shepherd has read, edited, critiqued, and in general made this material sensible and useful for readers. Craig Polak stepped in when Dave was busy on other tasks. Carla Williams picked up loose ends, made sure faxes got faxed, and kept everything organized. John Lazar, Don Kumler, and Ken Silber researched, advised, and kept me at task. This book is the result of the effort of our whole organization. I may be the primary author, but I know for certain the project would not have been completed without their help and support.

Much of the adhesive that holds this book together has been supplied by my husband and colleague, Odin Westgaard. He has a talent for developing examples and illustrations.

THE PUBLISHER

Writers are notorious for complaining about their publishers. At times I have had similar feelings. But not for this book. The people at Quality Resources, especially Cindy Tokumitsu, have been understanding, supportive, and very accessible at all stages of the project. I am grateful for their help and their ability to be flexible about schedules.

A SPECIAL BIRD

The peregrine falcon.

INTRODUCTION

The training function plays an important role in helping organizations win a Baldrige Award, get ISO certification, or deploy TQM. For example, employees at all levels attend training to learn what quality means, what role they are expected to play, how to work together in teams, how to chart their processes, how to define and apply measures, and how to achieve continuous improvement.

However, this book is not about getting ISO certification or deploying TQM. Likewise, this book does not deal with how to win the Baldrige Award. This book instead takes the position that training managers and practitioners should apply the Baldrige, ISO, and TQM standards to the management of the training function. It explains how they should:

- Set standards for the work they produce or commission.

- Evaluate the function's effectiveness.

- Select and work with suppliers and vendors.

- Determine the type of relationship they want with their customers and suppliers.

- Define what role they will play in their larger organization or in the marketplace.

By applying the guidelines in this book the training function is better able to support the larger organization. Training can be instrumental in helping its organization join the quality movement. More important, it can help executive management optimize its investment in people throughout the company.

The book offers training managers and practitioners:

- A process to evaluate how well their training exemplifies the quality principles.

- Guidelines and tools to use to apply quality principles to all phases of training and the development process.

- Examples of what other training functions are doing to embrace the quality principles of customer satisfaction, measurable results, leadership, and fact-based management.

- Specific advice training managers can use to move from a subservient support function to equal partnership with dominant players such as operations, marketing, and accounting.

The book is divided into six chapters. Each can (and probably should) be considered an important building block for the training function of the future.

Chapter One explores the need for, and provides guidance in, developing a business case for the training function. This chapter builds the case that training managers and professionals must play a leadership role in the quality movement by applying the quality concepts to the running of the training department.

Chapter Two is dedicated to the leadership training must exhibit to assume its new role. Here you will find what the role of leadership is, what it requires, and how it can be achieved. You will learn how this role can be demonstrated within the training function and the larger client organization.

Chapter Three explores what training has to do before it can be considered a strategic business partner (operate as a well-run business and be competitive). This chapter is based on the premise that the training function competes for resources. It explains how to determine if you (the function or a training professional) are competitive in the marketplace. Before the training function can be given peer status, it must define itself and prove its worth in the same ways the other business functions use.

Chapter Four explores how to describe and document current processes used to assess and describe needs; design and develop training services; select and carry out a delivery method; and measure and evaluate outcomes. Processes are the heart of the training function. If they are appropriate, efficient, and effective, then the function can be viable and can support the movement toward quality.

Chapter Five explains what the Baldrige, ISO, and International Board of Standards for Training, Performance, andInstruction (IBSTPI) standards are and why they are important. It details the training standards that are available, how these standards came about, ways they can be used, and how they support the Baldrige Award, ISO certification, and other quality efforts.

Chapter Six is about measuring success. It brings evaluation into focus and supplies tools for making good evaluation happen. The chapter emphasizes the importance of applying the best measures at the right time.

Descriptions of the criteria relating to training from Baldrige, ISO, IBSTPI, and pioneering work done in England and elsewhere are provided in the Appendixes.

WHAT DOES QUALITY MEAN FOR TRAINING MANAGERS?

Quality requires two basic conditions. First, attainable, consistent, and well-established standards must be set, otherwise there is nothing against which to measure success. Second, quality depends on the skills and dedication of the producers.

Quality is applicable, as a concept, to almost any endeavor or product including training, because quality training must meet established standards. In addition, it is provided by people who must have the necesssary skills and knowledge.

Quality is the characteristic of a product, service, or process that enables it to consistently meet both internal (producer) and external (customer) expectations. Quality is measured by comparing a finished product or service to the specifications set by its designers. Quality is measured by customer satisfaction, continuous improvement, efficient use of resources, and uniformity of performance. The final imperative is acceptance by customers.

Quality training is valid, reliable, responsive, and meets (if not exceeds) customer expectations. Specifically, when quality has been achieved:

- Courses perform as expected over repeated offerings; they are reliable.
- Different courses offered by the same vendor meet equivalent standards; they perform as advertised.
- Instructors deliver programs with consistent results.
- The process used to design, develop, and deliver courses is efficient in terms of time and dollars; the process uses resources well.
- Customers can speak to the efficacy of learning opportunities; customers acknowledge training's value.
- Training personnel can speak of the business implications of training's products and services.
- Customers cite the training function as the supplier of choice. Training is competitive and adds value in the eyes of the customer.
- The training function monitors its processes and actively seeks ways to eliminate rework, reduce cycle time, and be responsive. There is continuous improvement of processes and output.

THE TRAINING FUNCTION AND LEARNING

The training function cannot achieve quality without learning. It must learn to better manage and improve its internal processes. It must also learn its customers' mission, goals, processes, products, and market pressures. Learning at this level happens when training managers and practitioners purposely acquire, validate, and appropriately use information. This does not come through happenstance. This learning occurs when you prepare and take the initiative to inquire, experiment, and test.

THE IMPACT OF THE TRAINING FUNCTION

This book focuses on the quality principles the training function can and should adopt. It is about the role training managers have in the new order. It is supported by three important ideas:

1. None of the aspects of the quality movement can be implemented without training. The Baldrige criteria, the ISO qualification standards, and other quality applications depend on appropriate, efficient, and effective training.

2. Quality can't happen in isolation. For a product or service to maintain consistent quality it must be supported by qualified agencies.

3. The customer is more important than ever before. Making customers happy by meeting their expectations and gaining their acceptance is critical. And the task is never finished because customer expectations escalate as they learn.

1
TRAINING AND QUALITY

> *A peregrine falcon is beautiful in appearance and flight. Perhaps, more important, it is an excellent hunter. Leaving its nest high on a cliff face, it soars, dips, and dives with an ease that belies the power and precision of its hunt. It preys upon other birds—usually ducks—that it attacks in midair.*
>
> *Twenty years ago, peregrines were sorely pressed to survive. They were high on the endangered species list. The reasons were fairly easy to understand. Most of them stemmed from the encroachment of people on the wilderness.*
>
> *Peregrines are still threatened, but not nearly as much as in the past. Why? They have been introduced to city living. Instead of cliffs, they nest on the perpendicular sides of skyscrapers. Instead of ducks, they eat pigeons. It is a major paradigm shift for the peregrines and an important lesson. The falcons' world had changed, never to return to its old form. Instead of trying to redevelop the wilderness setting, a virtual impossibility, conservationists have introduced the falcons to another similar environment where the birds have found plentitude. There are hundreds of skyscrapers upon which to nest. There are thousands of pigeons to eat. It is a good place for the peregrines.*

It's an interesting story, isn't it? But what does it have to do with training, and, in particular, quality criteria, standards, and principles as applied to training?

Training, too, has encountered a change of major proportions. Like peregrines, if we training professionals want to survive, we must learn new ways, discover new sources of sustenance, and find new homes. We cannot ignore the quality movement. It's here to stay and we must become part of it. This book is about how to do it elegantly. This book is about how to follow the paradigm shift, take advantage of it, even lead it, and emerge healthier and more respected than ever.

Part of the paradigm shift is to acknowledge customers. Satisfied customers are critical. So who are they? They are the line managers of key business units. Why line managers? Because they direct their departments and organizations. Their work units must provide cost-competitive, value-added services if they are to survive, and their survival depends on well designed processes, the latest technology, and competent workers. They are the customers, managers who decide whether they and their workers want your training.

Quality is judged by customer satisfaction and product performance. Customer satisfaction is achieved through knowing customers, involving customers in defining products and measures of success, and informing customers about changes made to accommodate their requirements. Product performance for training means that courses and other training modes consistently accomplish desired results. Consistent performance is measured through reduced variability (i.e., people and products perform reliably and with equal effect). For training, consistent performance shows up as follows:

- An instructional program or product performs as expected for repeated offerings. It is reliable.

- Different instructional programs and products offered by the same vendor or department meet equivalent standards. Different programs can be trusted to perform as advertised. They are equivalent.

- An instructor delivers programs with consistent results and therefore is reliable.

- Different instructors possess the same level of competence and can be substituted one for another. They are equivalent.

This chapter builds a business case for why quality principles should be applied to all aspects of training—its management, processes, products, and services. It describes:

- The quality movement and its impact on training.
- Social, economic, and regulatory pressures and their impact on training.
- Competition and its impact on training.
- The range of services being provided by training functions.

Chapter 1 also introduces the format used throughout the book:

- Present information and examples.
- Provide questions to assess how well quality principles are applied now.
- Offer suggestions on how to proceed or what to do.
- Suggest references for more information on specific topics.

THE BUSINESS CASE

The training function in most organizations doesn't consider itself as a business. True, most of us in training know we must do our best to help our companies succeed. But are we businesses? Are we organized, motivated, and managed like a real, honest-to-goodness business? In a word, no. Many training departments are like caterers. They provide training on demand. They develop goodies for their clients.

Does your training department have its own mission and vision? Do you develop budgets based on anticipated revenues? Do you market your products and services? Perhaps it isn't fair to ask such questions. After all, chances are you haven't been empowered to act as a business. To become empowered you must convince decision makers that it makes good business sense for your department to be more businesslike. Without this vital first step, attempts to provide quality training are difficult.

This chapter focuses on some thinking that you can use to move in the right direction. It will help your training function become a viable business and take its rightful place as a critical part of the movement toward quality in your company. There are various issues that make up the business case. In the foreground is the quality movement. In the background are the changes in the social, economic, and business environments. Let's first look at the quality movement.

THE QUALITY MOVEMENT

In the 1980s, to become more competitive in global markets, U.S. business began to turn its attention to the quality movement. Two hallmarks of that movement—The Malcolm Baldrige National Quality Award and ISO 9000 quality standards—continue to stimulate conversations about what quality is, who is the judge, how it is measured, and its implication for profitability. The conversations in turn deepen the understanding of whose job quality is and how to meet quality standards. Figure 1.1 gives a brief overview of the Baldrige Award.

The Baldrige Award has had great influence nationally, but ISO 9000 has international influence. Figure 1.2 gives a brief synopsis of ISO 9000 quality standards.

Together, Baldrige and ISO redefined quality. Baldrige challenged the old paradigm that quality control experts were the only entitled judges. Baldrige legitimized "the voice of the customer" and positioned customer satisfaction as the ultimate measure. ISO dared to demand that people be qualified to do the jobs they are doing by training, experience, or examination whether they be employees, management, or suppliers. Both moved quality beyond the manufacturing function and expanded it to include the whole organization. Both demand that management provides leadership. Both require

FIGURE 1.1. Malcolm Baldrige
National Quality Award

The Malcolm Baldrige Award is named after a former U.S. Secretary of Commerce. The award was established by the Malcolm Baldrige National Quality Improvement Act signed by President Ronald Reagan in 1987. The act calls for a national quality award with guidelines for organizations to evaluate their quality efforts. The act stipulates no more than two awards per category (manufacturing, service, and small business) be given per year.

The National Institute of Standards and Technology (NIST) administers the award. Judges are quality experts from business, academia, and consulting firms. Judges assign up to 1,000 points in four broad categories:

1. Leadership (10% of the points).
2. Deployment (42% of the points) through:
    A. Fact based management.
    B. Strategic quality planning.
    C. Human resource planning.
    D. Quality assurance.
3. Quality results (18% of the points).
4. Customer satisfaction (30% of the points).

The application process requires companies to answer a questionnaire covering 32 areas. Companies whose applications receive a high enough score are scheduled for an on-site evaluation. A team of examiners visits the company's site for several days. The purpose of the visit is to validate the information presented in the application. Judges then review the results of the visits and select winners.

Companies can choose how they accomplish their quality goals. Judges look for processes that achieve valid and reliable results that meet customer needs. Judges pay particular attention to:

- Quality programs that are customer focused.
- A high level of employee involvement.
- Management taking a leadership role.

FIGURE 1.2. ISO 9000 Quality Standards

International Standards Organization (ISO) Certification Registration was created in 1946 by 91 member nations to simplify the exchange of goods and services through the use of uniform standards. In 1987, ISO created the ISO 9000 quality standards for the design, manufacturing, inspection, packaging, and marketing of goods and services.

The ISO 9000 quality standards are divided into five sections. The first section, 9000, contains the general guidelines and definitions of terms referred to in the other standards. The last section, 9004, contains the standards for managing and auditing a quality system. The remaining sections refer to the three levels of certification. ISO 9001 is the most comprehensive. It was originally designed for engineering, construction, and manufacturing facilities engaged in the design, development, production, installation, and servicing of products. It contains all the standards. ISO 9002 is less comprehensive. It was originally designed for such processing industries as chemicals, food, and pharmaceuticals. ISO 9003 is the least comprehensive. It was originally designed for facilities dedicated to inspecting and testing products. What is significant to training is that all three levels have the same standards for training.

To apply for ISO 9000 certification, firms must implement a quality management system and submit to a third party audit. ISO uses an accredited third party, called a registrar, to confirm the design, implementation, documentation, and effectiveness of the quality system. Registrars are accredited by the National Institute of Standards. The American National Standards Institute (ANSI) is the accredited registrar in the U.S.

---

documentation, well-defined measures, and continuous process improvement. They are changing the way companies are managed and structured. They are changing relationships among management, employees, and suppliers.

Baldrige and ISO require people throughout an organization to learn the organization's mission, vision, strategic goals, and standards. This has led to "a systematic change in the organization, requiring new ways of thinking, organizing, leading, collaborating, and working, in every function and process of the business."[1] In short, they are changing the way work gets done.

The number of companies adopting quality principles—Baldrige in particular—is increasing. The number of ISO registrations is also increasing, going from 222 during the first quarter of 1992 to 1,804 per quarter by the end of 1993. The primary reason U.S. companies go after ISO certification is to meet customer expectations.[2] When U.S. and Canadian ISO-certified companies were asked the top three reasons for seeking ISO registration, 60% listed customer demand and expectations in the top three, 62% pointed to quality benefits, and 60% indicated market advantage.[3]

ISO registration also has a domino effect. ISO-certified companies ask their suppliers to become certified, and those suppliers in turn ask their suppliers to become certified. Why? Companies are finding that insisting on quality from suppliers results in improved customer satisfaction, reduced costs, reduced defects and errors, and improved service.[4] This chain effect has enabled companies to control not only output but input as well.

To comply with the requirements for the Baldrige Award or ISO certification, companies have developed many types of quality initiatives. The most frequently mentioned initiatives are, "training, self-assessment against the Baldrige criteria, customer focus, participation, empowerment, planning, leadership, and teamwork."[5] Even those companies not seeking the Baldrige Award or ISO certification have adopted many of these initiatives in order to stay competitive themselves. A closer look at four of these initiatives is taken in the following sections.

---

[1] A. Hiam, "Does Quality Work? A Review of Relevant Studies," *Report Number 1043*, The Conference Board (New York, 1993).
[2] *Quality Systems Update: Special Report*, CEEM Information Services (Fairfax VA, Fall 1993).
[3] *Quality Systems Update: Special Report*.
[4] Hiam.
[5] Hiam.

Training. Training plays a key role in the quality mandate. Baldrige criteria and ISO certification require training to:

- Impart quality principles.
- Create effective, self-managed teams.
- Deploy the organization's vision, mission, and goals.
- Validate that employees and vendors have the skills and knowledge they need to produce defect-free products and world-class service.

ISO certification, in particular, has a double impact on training:

1. ISO has standards for the training function.
2. Documented training of various staff groups will be needed to achieve ISO certification.

As a result, companies implementing quality initiatives have "made significant investments in training, reorganization, and work-flow redesign."[6] The most common training topics arising out of these initiatives include: leadership, team building, problem solving, quality improvement, managing change, and creativity.[7]

Self-Assessment. The majority of companies request Baldrige applications to use as a self-assessment and a model to create customized quality programs. Baldrige has done U.S. businesses a great, unanticipated service. The application procedure is an excellent tool for evaluating quality efforts and deploying quality principles.

Participation. Participation refers to how the vision, mission, and goals are deployed. The organization's vision, mission, and goals are no longer just for management. As decision

---
[6] Hiam.
[7] Hiam.

making moves down through the ranks of an organization and all employees become empowered to make important decisions, each employee must know the vision because each is accountable to achieve it.

Traditional organizations had formal mission statements. However, they were often used by managers to justify their decisions when challenged. Non-managers were asked to accept the manager's interpretation of the mission on faith. The new mandate is for every employee to understand the mission well enough to organize their work around it. Furthermore, an employee who observes actions incongruent with the mission is empowered to do something about it. Part of training's role is to expose employees' discomfort with this radical change in employee roles and responsibilities and to create opportunities for them to voice their concerns. Another part of training's role is to facilitate an organization's ability to make wise decisions by using a sound decision-making process and to teach how to access information. This means the wraps come off information and information is made accessible.

Self-Managed Teams. Baldrige and ISO promote the use of teams and empowered workers. Teams that operate efficiently and effectively don't just happen. The internal and external interactions between team members and others are too complex and fast-paced for trial-and-error development. Roles, responsibilities, and the basic nature of the work are different for self-managed teams than for traditional employee–supervisor roles or even for traditional teams directed by managers. The self-managed concept is foreign to traditional operations. Training is expected to teach people to understand what self-managed means, how it helps them improve the way they do things, and how to adapt to working in a self-managed mode.

Management is discovering that workers in self-managed teams require new skills in team development, communications, and problem solving. They must also monitor the efficacy of the processes they use to do work. Evaluating processes requires higher-level analytical skills. As a result of these demands, the search is on for ways to help each worker be:

- Individually responsible for productivity.
- Empowered to make decisions affecting self, team, and organization.
- An effective cross-functional team member.
- Fact-based in decision making.

Essentially, training is required to give everyone the skills needed to function as empowered workers who can troubleshoot, meet customer needs, identify defects, improve processes, collaborate effectively, and apply process controls. Traditionally, these functions were wholly vested in management. In both the Baldrige and ISO programs, they have moved into the work force.

Organizational effectiveness has greatly improved through the implementation of quality initiatives. In addition, the Baldrige Award and ISO certification have compelled organizations to validate the competency of employees and vendors and to train managers in the quality principles. The criteria force managers to describe and measure processes and performance.

Validate Employee and Vendor Competence. Assuring everyone has the knowledge and skills to produce quality goods or services is a check on organizational competence. An organization can be competent only if each individual employee is competent. Validation of competency requires at least four kinds of investigation:

1. Employees must be competent. They must be skillful at doing what they are supposed to be doing. They must contribute to the organization's productivity.

2. Employees must be working on appropriate tasks at appropriate times. Their activities must contribute, at the moment, to achieving the organization's mission.

3. Employees must produce appropriate goods and services on time and within budget. They must contribute to the organization's cost-competitiveness.

4. The goods or services people produce must meet the quality standards. They must contribute to the organization's ability to deliver quality products its customers want.

These requirements affect suppliers as well. Managers must expect their suppliers to:

- Adopt the same level of rigor required by their internal quality standards.
- Measure results and compare them with a worthy benchmark.
- Account for cycle time.
- Seek ways to eliminate waste from processes.
- Consistently deliver products that meet quality standards.
- Produce more with less, faster.

Training in Quality Principles. Quality training programs give line managers (i.e., major customers and consumers of training) the tools and skills to measure processes, inputs, and outputs; chart variance; identify process disconnects; do cause analysis and generate alternative solutions. It is a fairly safe assumption that senior managers expect training managers to learn to do the same as managers in any other organizational function.

The effect of the quality movement on the changing face of business is clear and evident. Many organizational changes can be directy attributed to companies who have been influenced by the Baldrige Award or ISO 9000. While the quality movement—and its offspring, such as reengineering—have been in the foreground influencing business, there have been other pressures at work in the background.

BACKGROUND ISSUES: SOCIAL, ECONOMIC, AND BUSINESS PRESSURES

Social Pressures

Diversity. The work force is increasingly more diverse and companies rely on training to help prepare and integrate new workers. Training is helping businesses learn to work effectively in a diverse environment. Although global businesses face even stronger pressures, domestic companies are recognizing that the profile of the average U.S. worker is changing. In 1991 approximately 40% of U.S. businesses saw diversity as a priority issue. Another 33% were beginning to address the issue.

Consider these changes:

- The work force is becoming older and is getting more training. Workers between the ages of 35 and 44 received 71% more skill improvement training in 1991 than in 1983. This age group receives the biggest share of skill-improvement training, 31%.[8]

- The work force is becoming a more heterogeneous mix of races, ethnicities, and cultures. Of the six million children from age five to 17 who speak a language other than English in their homes, one million speak English

---

[8] *Cultural Diversity Survey* #5 (American Society for Training and Development, 1991).

poorly or not at all.[9] According to Thomas Exter, within 15 years more than half of the nation's new workers will be minorities.[10] Hispanic employees saw a 120% increase in the amount of skill-improvement training received from 1983 to 1991. Blacks saw a 59% increase in this training, and whites saw a 36% increase.[11]

- Continuing efforts to integrate people with disabilities into the mainstream have enabled more people with physical limitations to work in areas where they have been denied access in the past. Training's role in this effort is in the design of performance support systems.

- Women are 45.3% of the work force and 48% more women reported receiving training in 1991 than in 1983.[12]

The impact on training. Training's customers don't all look or think the same way. Training must be inclusive of everyone. That requires a sharper audience analysis to define the breadth of learner needs and points of view. For example, the advertising industry's television commercials have become much more inclusive in the last several years. No longer do only men drive the cars in automobile commercials. No longer do only women hold babies or a bottle of floor cleaner. Not all shoppers are walking in a retail setting—some are in wheelchairs.

Training has the same accountability. Training must accurately portray the diversity of the work force. Trainers must make sure training materials are inclusive of the diversity of the entire work force. Examples must be balanced and avoid stereotypes. Instructional strategies and tactics must be inclu-

---

[9] A. Carnivale and E. Carnivale, "Growth Patterns in Workplace Training," *Training and Development Journal* (May 1994).
[10] T. G. Exter, *The Declining Majority, American Demographics* (New York: Dow Jones and Co., 1993).
[11] Exter.
[12] Carnivale and Carnivale.

sive. For example, a small group discussion may not work well when participants have widely different proficiency in English or have cultural expectations that the leader's role is to dominate discussion. Diversity issues require training to provide even more performance support systems, self-paced instruction, and job aids. In addition to ensuring that training programs address the diverse needs of an organization's employees, new cultural diversity programs are created to enable all employees to learn how to function effectively together. Internal training staff are the primary providers of cultural diversity programs. The majority of the programs offered are customized rather than off-the-shelf.[13]

Training staff are put in an unfamiliar and sometimes awkward position. Baldrige and ISO demand consistent, reliable quality efforts. Every service or product is expected to achieve the standards set. This requires consummate teamwork not only within the training function but also with the people the training function services, their clients and customers. Teamwork with diverse elements sounds difficult. In some organizations it is. Those, fortunately, are exceptions rather than the rule.

It makes sense for training to take a leadership role in this regard. Training professionals in most organizations are aware of the needs of the diverse elements within the organization. They must often balance those needs against the goals and focus of disparate forces. For example, there may be an organizational mandate to hire more minority ethnic groups such as Asian Americans or Latinos. The new hires will inject new cultural values and mores into the existing cultural matrix of the organization. To ignore the situation is to court trouble. Management, however, is usually ill equipped to deal with this kind of situation even though it is common in almost every organization.

The training function is the obvious answer. Quality depends on everyone working together with the same focus and

---

[13] Carnivale and Carnivale.

direction. Baldrige and ISO standards apply to everyone in the organization. Only the training function can simultaneously affect all these diverse elements without special actions. When training is included in the effort to accomodate diversity a wide variety of activities can be enhanced. Problem solving, morale, and skill development are obvious arenas where an impact can be made fairly quickly without undue expense.

Process Redesign and an Underskilled Work Force. In addition to training, businesses are investing in the redesign of their processes. Global competition is forcing companies to redefine jobs and processes. They are reengineering their processes to improve quality and competitiveness. One outcome is a clearer understanding of core business activities.

Companies are reducing work forces while expecting employees to be more productive than ever. Companies that downsize don't expect to produce less. On the contrary, they hope to produce more at higher levels of quality with better technology, streamlined processes, and fewer layers of management.

As a result, some employees are displaced because they are redundant or not at the right skill level. At the same time, the education level of workers is moving towards opposite ends of the spectrum. Workers are either "over educated" or without basic work skills.

We've all heard stories about people with Ph.D.s working as cab drivers. Although many people with doctorates are not displaced, many others are. People take jobs for security or to satisfy basic needs. If those jobs are below their capabilities, they often become bored or disinterested and, subsequently, perform poorly. Jobs that are mechanical in nature such as assembly line work are often hard to fill with well-motivated people for the same reasons. A much larger problem lies with those whose skills are not sufficient, either because they never adequately learned them or the skills have become obsolete because of technological advances.

...ct on training. Both the dependence on technology ...henomenon of underskilled, displaced workers are ...ew demands for training. Workers need training in tec... , life, and basic skills. Even well-educated workers need retraining to stay up to date in their professional discipline.

Basically, a business consistently delivers high quality because its workers are competent. The best equipment and plant cannot produce top quality products without competent workers. Competency is learned. The function that supports competency is training. Training is expected to identify and develop compentency at the organizational, departmental, and individual levels. Training is expected to prepare workers for new jobs and processes. Training is needed to retool the work force.

New Employer Relationships and Alliances. Businesses are entering into new alliances with employees, suppliers, customers, and competitors to achieve their goals in cost-competitive ways. Relationships between employers and workers are being redefined. Companies are outsourcing all but their core business functions. For instance, some companies are outsourcing accounts payable and information systems functions. Temporary workers, part-time contractors, and independent workers are the fastest growing employment categories. Service industries now generate 9 out of 10 new jobs. Companies are setting standards for their suppliers and asking that suppliers be certified.

The impact on training. Businesses look to training to help qualify workers to operate effectively in these new arrangements. They look to training to help develop the interpersonal, teamwork, and negotiation skills to support these alliances.

The Importance of Training. In short, because of a changing work force, process redesign, and new work relationships, training is no longer a supplementary activity but is essential for survival. All oganizations must maintain or increase productivity. They can't wait for an individual to return to school to learn new skills. They can't rely on apprentice systems. On-the-job learning no longer suffices to provide the competency needed for quality work; the tutors tend to pass on obsolete skill sets.

No one, no matter how talented or educated, can hope to maintain competent performance over time in today's business world. One analysis of training data from the U.S. Census Bureau, U.S. Department of Labor, and the Small Business Administration showed that formal company training increased 45% from 1983 to 1991; in 1983, 35% of workers reported receiving skill-improvement training and in 1991, 41% of workers reported this kind of training.[14] According to one report: "Estimates of employer investment in workplace training hover around $210 billion annually. About $30 billion, or 1 or 2% of employers' payrolls, is estimated to be spent on formal training, while another $180 billion annually is invested in informal, or on-the-job, training."[15]

One million student hours of training a year is an accepted fact for many companies. Training is no longer an exceptional occurrence for most workers. It's something they count on, look forward to.[16] Some business leaders mandate either a minimum number of hours a year or a percentage of payroll be budgeted to training. For example:

- Saturn requires every employee get a minimum of 92 hours of training annually.

- Business units within AT&T require 40 hours of training annually.

- Allstate Business Insurance budgets two weeks of training for every employee.

---

[14] Carnivale and Carnivale.
[15] A. P. Carnevale, L. J. Gainer, and J. Villet, *Training in America: The Organization and Strategic Role of Training* (San Francisco: Jossey-Bass, 1990), p. 23.
[16] Carnivale and Carnivale.

Training is something they have to have to remain productive. At the same time, it is estimated that 42% of the work force or 50 million workers will get the training they need.[17]

Economic Pressures

In addition to the social pressures influencing the training function, economic pressures also are changing the face of the training function. This is being felt mostly in the area of competition.

New Competitors. The amount of money being spent on training is attracting new players. Those new players may not know how to make learning happen. They may not know what it takes to improve performance. Whether these new vendors are competent, customers have alternatives they've never had before.

Advertising agencies are one example of the new competition. For instance, one major advertising firm, hired by a sales and marketing executive to put on sales conferences to launch new products and annual promotions, found out during the conference what its large corporate client spent on training. It created a strategic initiative to be the sole source for training for those clients. The advertising account executive stated, "We know the training game. It's having breakout [sessions]."

Another advertising agency was approached by a manufacturer of health care equipment. The company went to the advertising agency to develop a workshop to teach nurses how to deal more effectively with AIDS patients and how to calibrate its equipment. Why did the manufacturer consider an ad agency? The senior manager had experience dealing with advertising account executives. He knew advertising agencies were good at packaging messages. He didn't even think of his training department. He believed the ad agency's

---

[17] "The Training Gap," *Training and Development Journal* (March 1991), pp. 9–10.

business is communications while training's business is learning and performance improvement. Many customers don't know the difference.

Direct Competition. The number of competitors for the training dollar has increased because the market is good and growing. With that kind of incentive it's no wonder the competition is strong. It will probably get stronger as more people learn about the opportunities.

External vendors are now selling products and services to customers previously served only by internal training resources. Vendors are aggressively marketing to line managers through direct mail and sales calls, bypassing the training function. Some are willing to lower their price to gain access to new customers.

Major accounting firms such as Arthur Andersen & Co. offer consulting and training services, and large companies are setting up their training departments to operate as separate businesses. For example, IBM spun off its training department and created Skill Dynamics. Even training departments not operating as separate businesses are beginning to market their programs and services outside of their organization. Florida Power and Light, PacBell, and others have invested significant amounts of money in their training. They are looking to new markets to recover those costs.

University schools of management are aggressively going after the corporate management development market. Some integrate short-term "in-residence" programs with on-the-job assignments. Others offer weekend college. Business and trade associations are entering into special alliances with community colleges. Burlington Northern Railroad, for example, has a special relationship with Johnson County Community College, and Whirlpool uses the community college in Benton Harbor, Michigan. Even trade associations are moving their headquarters on campus. In addition, professors either consult part-time or leave for higher compensation as outside training consultants.

Finally, many experienced trainers found their jobs were eliminated, yet their services were still in demand. They became independent consultants to the very organizations that had formerly employed them. The company, therefore, retains the services of capable people while reducing its overhead.

The impact on training. The consequences to the training function are painfully clear. There are thousands more training vendors entering the market who can produce quality products and services. They can offer their services at competitive prices. Customers can demand a full range of training services and have several options to fulfill those demands.

Business Pressures

Related to the economic pressures are the business pressures forcing change upon the training function. Where economic pressures involve competition, business pressures have to do with "working smart"—doing what makes good business sense, holding people accountable, and other similar activities.

Return on Training Investment. Although training is one of the fastest growing industries, it is considered the most expensive way to transfer new skills. However, training is invariably required to create and maintain a work force capable of building quality products and services. The net effect is that training is less likely to be seen as a "nice-to-have," and more likely to be seen as a critical investment in the employees of an organization. As with any investment, business leaders expect a return on this investment such as increased productivity, market share, and profitability. Customers expect training to take on the same degree of rigor they are demanding of core business functions. They expect training departments, vendors, and consultants to add value and to be able to prove the results of their products and services. They want the evidence of reduced cycle time and decreased variability in products sold to customers.

Training is no longer being issued a blanket exemption from cost-competitiveness. On the contrary, internal training departments are expected to be price competitive. However, internal training functions can have a tremendous competitive advantage over outside vendors if their manager understands their organization's business, customers, processes, and performance. For instance, companies are aware that quality is built internally, and who understands the organization better—an external vendor or the internal training function? To take advantage of this opportunity, though, an internal training function can respond to these quality-enhancement opportunities if it:

- Develops its own skills in team-building, facilitation, and evaluation.
- Develops a quality process and standards specific to training's needs.
- Evaluates and improves its own processes and products.

For example, many training departments have traditionally maintained full video production facilities to produce their own videos. However, video production is technology-driven. To stay competitive, production equipment must be constantly upgraded. A company shouldn't be surprised to hear a training manager say: "We can no longer afford to keep our own video production facility at a state-of-the-art level needed to stay cost-competitive. Outsourcing video production to a company that is willing to invest in the technology will be more cost-effective." The message: to add value, a training function must be aggressive about managing training resources and insuring good return on investment. It should identify those services that are best outsourced, set standards for all products and processes, and put in place ways to improve what it does.

Businesses wants a clear return on its investment in training in terms of improved performance. Business expects a training department to:

- Know what its customers want and need.
- Know what services businesses are buying, how much they are spending, and what results they are getting.
- Do competitive analyses and make use of processes that result in a competitive advantage.
- Know its costs and prove its services add value.

This is not difficult. The story of Sandia National Laboratories in Figure 1.3 explains how getting a return on an investment is possible.

The Sandia training team showed a $200 return for every dollar invested. That's a remarkable return for a business, let alone a training function. It took a commitment on Sandia's part to measure its return on investment, but it turned out well. When the organization demands accountability from the training function, training can respond.

SUMMARY

Training has a role to play in helping organizations be competitive. The role has always been there, but it's a much larger role today than ever before, and it's expanding. Figure 1.4 lists some of the services that training departments now provide.

Most organizations now hold the training operation accountable for learning and performance improvement. On-the-job training and learn-as-you-go are tactics that no longer work. So, formal training has become necessary. Whether the internal training function is one person, or a large department with hundreds of professionals, the mandate is the same: "Bring our people up to speed (i.e., complete competence) as quickly as possible and do it for less money."

There is little or no time nor funds available for long, involved development projects. There is little support for ineffective training in which learners sit passively in classrooms. What business organizations want from training corresponds

FIGURE 1.3. The Sandia Story

Sandia National Laboratories (owned by the U.S. Department of Energy and managed by Martin Marietta) has made a public commitment to quality. The President's Quality Award is given annually to "Sandians and Sandia teams who make outstanding contributions to the success of the Laboratories and who accelerate Sandia's development in becoming a national leader in quality and quality progress." There are three levels of awards and no limit to the number of awards given. The award is modeled after the Baldrige Award and ISO certification. The 1993 process was structured as follows:

- **Customers and stakeholders** (*100 points*). Identification of the customer and the relationship of the activity to Sandia's strategies and goals.

- **Process** (*450 points*). Identification of customer expectations, management of customer relations, production processes, goods and services of suppliers, methods to evaluate customer satisfaction, and improvement of key processes.

- **Results** (450 points). Results of the activity and customer satisfaction.

In 1994, Charline Seyfer, program manager, received one of the awards and said, "Development of a training course may, at first blush, appear relatively inconsequential when compared to placing satellites on the moon; however, the **impact** of a **highly effective, student-time efficient course** that **contributes to improved** employee **performance** and provides a **return on the training investment** is long term and far reaching . . ." The training team was able to link the course development project to Sandia's long term strategies and goals. The team documented the instructional analysis, design, development, delivery, and evaluation processes it used, how it confirmed the customer need, aligned the product with that need, and qualified its vendor partner. The team was able to prove customer satisfaction (evaluations ranged from 4.5 and 5.0 on a 5 point scale). Six months after the course the team conducted level 3 (i.e., application to the job) and level 4 (i.e., business results) evaluation. The evaluation confirmed the procedures and guidelines presented in the course were still being used by attendees. It was able to show a positive return on investment: "For **every dollar** in development of the course Sandia received a $200 return."

Source: Charline Seyfer, Sandia National Laboratory

FIGURE 1.4. Scope of Training Services

1. *Needs assessment.* Discover specifically what the business customers want and need.
2. *Needs analysis.* Define what skills, knowledge, and support systems are required to meet the business need.
3. *Task analysis.* Specify what knowledge and skills are necessary (and supplementary) for the competent execution of a task.
4. *Instructional design.* Design instructional episodes that will deliver exactly what the customer wants.
5. *Instructional development.* Prepare materials, media, and processes learners can use to learn.
6. *Administer and deliver training programs.* Administer self-paced and computer-based programs, conduct workshops and seminars.
7. *Management of formal learning environments (corporate colleges and training facilities).* Provide a facility where anyone in the organization can go to learn what he or she wants to learn.
8. *Multimedia production centers.* Prepare professional materials specifically designed for the job to be done.
9. *Management of contract services.* Screen, select, and manage training vendors and other suppliers efficiently and effectively to develop or deliver programs.
10. *Recruitment and development of instructors.* Provide the best available practitioners for the specific needs of the customers.
11. *Maintain student records.* Track learners and maintain accurate, current records of their competence.
12. *Measure competence.* Provide tests and evaluation techniques to assess individual and corporate performance levels.
13. *Certify competence.* Provide a certification program for internal operations of all types.
14. *Provide documentation services.* Publish and preserve documents important to organizational aspirations of any sort.
15. *Market products and services.* Inform internal and external customers about what training offers, the cost/benefit of programs, and why securing these things from training is good business.

to what they want from their other suppliers of key products: just-in-time training that works, training designed for their specific circumstances, and training that is convenient, reliable, and competitive.

For a training function to become better at providing quality products and services it may require help. One organization trainers can turn to is the International Board of Standards for Training, Performance, and Instruction (IBSTPI), which is discussed in Figure 1.5.

IBSTPI saw a need for a certification program for training departments. Training departments are the agents responsible for implementing change in the pursuit of quality. Quality of processes, products, and services, as defined by ISO and Baldrige standards, can only be achieved through organizational learning. Here, learning means purposely acquiring and evaluating information. What is learned can then be applied to align one's products and services to support the host

FIGURE 1.5. The International Board of Standards for Training, Performance, and Instruction (IBSTPI)

> IBSTPI stands for the International Board of Standards for Training, Performance, and Instruction. It was founded by the Joint AECT/NSPI Task Force on Standards and Certification. Representatives from the American Educational Communications Technology (AECT) and the National Society for Performance and Instruction (NSPI) incorporated IBSTPI as a not-for-profit organization in 1986. Its mission is to:
>
>> Promote high standards of professional practice in training, performance, and instruction for the benefit of individual and organizational consumers through research, definition of competencies, measurement of competencies, education, and certification.
>
> IBSTPI is run by a volunteer governing board made up of representatives from business, academia, and independent practitioners. IBSTPI identified and validated standards for instructional designers, instructors, and training managers. IBSTPI certifies instructional designers, instructors, training departments, and training products. The certification programs integrate their standards with ISO's 9000 standards and the Baldrige Award criteria.
>
> *Source:* IBSTPI

organization's goals and customer needs. IBSTPI set about to develop standards and criteria that training departments could use to help them consistently provide quality products and services. IBSTPI's standards and assessment process are designed to help the training function:

- Exemplify leadership.
- Operate as a well run business.
- Qualify its people, products, and vendors through standards.
- Measure and evaluate its products and services.
- Improve its processes.

IBSTPI's standards are the basis for how the following chapters are organized.

Check Your Understanding

This chapter would be incomplete if there were no tool you could use to assess how your training effort stacks up. An easy way for you to discover this is to complete the questionnaire about your customers, services, and competition. Use the checklist that follows to determine how well you understand the pressures your customers are facing and what they expect of you. Check yes or no in the appropriate column. Use the comments column if you aren't sure and to record any remarks. You can use the checklist as a way to build your own quality system. It can serve as an agenda for staff, customer, and vendor meetings.

Question	Yes	No	Comment
1. Do you know how much money your training customers have or are spending for training associated with diversity issues?			
2. Do you know how much money your customers are spending on training to support quality initiatives (e.g., empowerment, facilitation, vision/mission)?			
3. Do you involve your customers in identifying their needs, defining their requirements, and agreeing on what basis your work will be evaluated?			
4. Is the training function involved in the quality effort in your organization? If so, how?			
5. Do you feel you've deployed quality principles throughout the training department?			
6. Have you read the books your customers are reading about such issues as quality, process design, teamwork, and continuous improvement?			
7. Can the training staff and contractors explain the difference between the Baldrige Award and ISO certification?			
8. Is training supporting your customers' efforts to redesign their processes?			

Question	Yes	No	Comment
9. Are your staff members empowered to act independently on behalf of their customers?			
10. Does your staff collaborate with customers? Do they work as teams either within training or with the customer?			
11. Does your vision include how you will provide the services your customers want?			
12. Is your vision for quality consistent with and supportive of the vision of the organization?			
13. Do you have measures for all the services your customers want (so you can determine their value)?			
14. Have you analyzed the processes used to deliver the services your customers want?			

What To Do Next

1. Arrange for the training staff to attend a program on quality basics. Many companies contract out for this service and exclude their own training departments. Arrange to get video tapes and materials describing the Baldrige Award and the ISO 9000 standards. Send away for them for your own use.

2. Meet with your key customers to discuss where they are in the quality process and how you might help them. Arrange for a "Training Day" and involve them

in setting your measures and defining your trainer, product, and vendor standards.

3. Meet with the human resources department to find out what it sees as the diversity issues in the company.

4. Check with the company librarian or purchasing staff to find out what books are being requested and purchased.

5. Survey your customers to find out what their financial commitment is to quality, process redesign, diversity, professional development, and tuition reimbursement. Not all companies centralize these functions. Line managers frequently have a lot of latitude when it comes to purchasing consulting services or buying training programs. Even if an expense goes through your budget, line management ultimately pays for it.

6. Benchmark with other training departments to see how well they are able to evaluate their products and services and qualify their people.

7. Ask to assign a representative to key quality and process redesign teams.

8. Ask your staff members what they understand is the training function's vision, mission, and key strategies. Compare their answers with your customers' views.

2
TRAINING AND LEADERSHIP

At two of several skyscrapers in downtown Chicago where peregrine falcons nest, an interesting contrast has developed. At both of the buildings the people who live and work there have discovered falcons aren't necessarily tidy creatures. In fact, the birds have a habit of devouring their prey on window sills and leaving bones, feathers, and other ugly remains.

The people in one skyscraper, repulsed by the remains, have demanded that the city government "do something about it." They insist the people who introduced the falcons to the city are responsible for ensuring the birds are well behaved. In short, they are unwilling to adjust to a new set of circumstances. They want someone else to take care of what they believe interferes with their ability to work and produce competently.

At the other skyscraper the people inside have placed little barriers around nests so falcon chicks don't plunge to their deaths. They have shuttered windows looking out on messy scenes. Not only have they adjusted to the new situation, they have provided leadership in humane ways to deal with their new neighbors. Other buildings are beginning to adopt their approach.

This example of leadership seems obvious. However, these heroes could much more easily have adopted an attitude similar to that of their neighbor. They could have complained and waited for someone else to act. Is your training function like that? Are you waiting for someone else to save your bacon? Guess what? It probably won't happen. Your potential saviors, like Chicago's city government, have other, more pressing matters to occupy their energies and thinking.

Like the people in the skyscraper, you have several advantages that are inherent to your operation. Consider your people, your developers, instructors, and support staff. Who are they? What can they contribute to the business? What skills do they bring to the organization, not just in training-related areas but in all aspects of its operations? If the staff is experienced and has been with the organization for several years, they know a great deal about almost every phase of your customers' work. Your trainers know how work is done. They know who does it and why those people are successful. They have witnessed changes and helped make them happen. In many cases, an organization's training function fits the definition of a cadre: "a nucleus or core group, especially trained personnel, able to assume control and to train others."[1]

A cadre—a group able to assume control and train others. Think that way about the training function in your organization. What single attribute of a cadre is most essential to its success? Leadership.

The purpose of Chapter 2 is to make the case for why a training function should take a leadership role in building and maintaining its organization's competency in a rapidly changing business environment. Training managers must become credible leaders. This chapter describes:

- What leadership is.
- How leadership shows up in training.

WHAT LEADERSHIP IS

What is leadership? Classically, a leader is someone who, first, knows where a group is supposed to go and how it should get there. Second, a leader is willing to take charge, to accept the responsibility of moving a group toward a goal. Third, a leader sets a good example for those who follow. This is the definition of leadership used in this chapter.

---

[1] *Webster's Ninth New Collegiate Dictionary* (Merriam-Webster: Springfield MA).

Leadership can be found in individuals, groups, and functions. Most often, leaders are such individuals as CEOs or politicians, who give direction and force. Groups often do the same thing. A hospital staff may provide leadership in certain kinds of health care. A military unit may be considered to exemplify readiness. The same is true of functions, though it's more difficult to exemplify. Consider marketing as a function: in healthy organizations it provides leadership. The marketing function in many organizations defines the direction of the organization's core business; marketing provides leadership. Rarely, however, does the training function provide leadership.

Training, particularly in organizations with quality programs, can take the lead, show the way, and provide a consistent source of leadership. Why? The reasons are embedded in the definition of leadership. Training meets the criteria in the first part of the definition: it knows where its customers are going and what competencies are required to achieve success. What it may lack is the other two parts of the definition: the resolve to be accountable for the results and the ability to set a good example.

THE BALDRIGE AWARD AND LEADERSHIP

Leadership makes up 10% of the Malcolm Baldrige Award's criteria. These criteria define leadership as the creation and sustaining of clear and visible quality values and a system to guide all activities toward those values. Characteristics of the system are involvement at all levels and behaviors that are supportive and collaborative.

What a Baldrige committee accepts as evidence of leadership is the deployment (i.e., infusion, integration, and congruency) of an organization's public position on quality as compared to what it does and how it does it. Simply put, leadership is people modeling what they say they value. For example:

- Senior executives are expected to be personally involved in setting goals, planning, and reviewing performance as it relates to quality.

- There has to be a method to communicate quality values to everyone in an organization. In addition to deployment, quality values must be evaluated and continuously improved. The implications for a training function in this regard are clear and extensive.

- The behaviors and accomplishments of leadership should be easily observed by everyone in an organization. Leaders are expected to exercise responsibilities in a systematic way and their ideas and efforts should be available for inspection by stakeholders.

- Training must not only deploy quality principles, it must apply them, and must measure their impact. Training is responsible for inculcating quality.

CLEAR AND VISIBLE VALUES

Leaders provide direction when they communicate their values. Values are high-level standards people use to judge whether something is good or bad. Values are extremely powerful influences but are very difficult to articulate. They are usually defined by contrast: "Now here at XYZ, we would never do (what our competitor does)!" Values define the uniqueness of an organization. They are systemic in nature; they distinguish an organization from others and establish the context of where an organization fits in the complete social, political, and economic environment. Although they are dynamic (i.e., they change), they are relational, because they cannot be separated and evaluated out of context.

When you tell people what your values are you make it clear what you stand for. Others can use their own values as guides for making decisions about you. Being clear about what you value helps others deal with you more effectively.

It also creates an atmosphere where people inside and outside an organization can deal with each other ethically and judiciously.

For example, the purchasing officers at McDonald's Corporation published their values (See Figure 2.1). The purchasing officers are the senior management team responsible for global supply-chain management. They are managers by title and position. They demonstrated leadership when they made their values public. These values tell others how to interact with them.

The function best positioned to articulate and explain the organization's values is training because it maintains constant contact with its customers and is skilled in analyzing how

FIGURE 2.1. Purchasing Officers' Commitment to Each Other

- We share a vision and support each other.
- We work together as true partners toward achieving our vision. Our success is shared.
- We respect and value our differences. The fact that we don't all think alike makes us a more powerful team.
- We make time for each other. Our work together is amongst our highest priorities.
- Business conflict is healthy when resolved through an agreed process. We will avoid personal conflict.
- Ideas belong to the group, not an individual.
- We are totally open with each other; no secrets. Others should know that talking with one of us is like talking to all of us.
- We consider the impact on each other before making any decisions and let people know that's our policy.
- The team speaks with one voice on major issues.
- We actively listen to each other.

Source: McDonald's Corp.

values drive performance. At the same time, training must be willing to make public statements about what it values and how it operates.

IBSTPI put together a statement of values it asks training functions to publicly commit to (See Figure 2.2). Whether in a purchasing or a training department, the bottom line is that leaders declare what they believe in.

A SYSTEM TO GUIDE ALL ACTIVITIES

Leaders not only have clear and visible values, they also make sure they have a clear system to be in alignment with the rest of the organization. The system is the alignment of the values, mission, vision, and business strategy. Mission statements tell people what business an organization is in. Vision statements tell them what the organization wants to become. The strategy tells people how the organization is going to live up to its mission and move towards its vision. When an organization backs up its mission and vision with a plan, it gives employees what they need to direct their own activites.

A vision is of little value without the will to make it real. Much of that depends on how forcefully the organization's

FIGURE 2.2. IBSTPI's **Value Statement**

One behalf of my organization, I have and will continue to:

- Honor the laws and regulations governing the areas where we provide learning and performance improvement services.
- Adhere to IBSTPI's standards.
- Adhere to accepted business practices.
- Conduct business in such a way as to support the rights and dignity of individuals.

vision is deployed by training. More important is training's will to define and deploy its own vision.

HOW LEADERSHIP SHOWS UP IN TRAINING

With a clearer understanding of what leadership is, let's consider how leadership shows up in the training function. Leadership shows up when people:

- Share their mission and vision.
- Influence others' behavior and beliefs.
- Use power.
- Contend with complex social settings where there is conflict over priorities, goals, and the perception of reality.
- Accept and invite evaluation.
- Drive learning.
- Act as stewards for the ideals of the organization.[2,3]

The rest of this chapter looks at these seven items and how they apply to the training function's leadership.

THE TRAINING MISSION

A mission statement should reflect the training function's driving force. Often training managers feel their job is to implement the organization's mission rather than develop one

---

[2] L. E. Shaller, *Concepts and Skills for Leaders: Getting Things Done* 8th ed. (New York: Abingdon Press), pp. 146–150.
[3] P. M. Senge, "The Leader's New Work: Building Learning Organizations," *Harvard Business Review* 32, no. 1 (1990).

for themselves. Although training's mission must support the organization's mission, a competitive training organization has a mission statement that declares what business it is in and what it delivers to what customers. Otherwise, the function is at a competitive disadvantage.

The full impact of this idea becomes clear with a look at what a mission statement is and what it means. A mission statement is the purpose and aims of the organization. An organization is the sum of the products and services it offers and the customers it serves. Its activities, resources, plans, competencies, structure, and decision making are directed toward those products and markets. The basic question is, "Why this business, these products, these markets, instead of others?" The mission addresses these issues:[4]

- What is training's driving force?
- Who are its customers? How will it capture and retain them?
- What are its customer requirements for the products and services it provides?
- How will it develop and maintain the competency demanded by its mission and its customers' requirements?

The issue of driving forces is an important one. Let's look more closely at that particular issue.

The Driving Force. The driving force is the single overriding premise on which the training department bases its decisions and actions. It is the overall purpose or reason for being. For example, is training's primary purpose to:

- Satisfy customer needs?

---
[4]J. Hale, *Strategic Planning Worksheets and Guidelines* (Western Springs IL: Hale Associates, 1994).

- Promote its current products?
- Leverage its internal capabilities?
- Get a return on investment?

Customer requirements. Customers are any group who all want the same things. The group can be defined by such criteria as department, job, or job level. For example, Walgreen's training and development department's sole customer base is the drug store. McDonald's training customers are the people who work in the restaurant or who manage the people who work in the restaurant. Customers' wants become the target of training's efforts. What they want dictates what training is offered.

If customer needs is the driving force, a trainer searches for what customers want now, what is driving *their* business opportunities, their problems, and their emerging needs. The focus is analysis of customers needs—now and in the future. Developing new products, now, is a trainer's bread and butter.

Many training functions, whether they realize it or not, are customer driven. They continually test the waters to anticipate what their customers are going to need and develop the capabilities and technology to provide those products and services. If they were to wait for customers to come to them for help, they would miss the opportunity to offer the product. Because of the rapid changes in business operations, it makes sense that more training functions are finding that their driving force is the customer. For example, a training function is customer driven when it develops training to meet the increasing diversity of the work place in age, education, and technical expertise.

In terms of quality, customers' needs would seem to be the only driving force to consider. However, for training, its current course (i.e., products) and internal capabilities are more likely to drive its decisions.

Promote products. A driving force may be dictated by the training products already in place or the services (e.g., vendor management, facilitation, assessment) currently offered. If this is the case, a training manager will want to look carefully at the products and services in place. Usually training's product line is set. It may improve or change in appearance, but its basic nature will remain. Just as automobile manufacturers will be in the car business during the next 20 years, training will continue to be in the teaching, people development, and performance improvement business. The challenge is getting customers to buy. If the product line is the driving force, the training department will want to focus on increasing its current customer base. It will want to find ways to improve and extend the life of its products. It will use its resources to develop, produce, promote, sell, deliver, and service its current product line to more customers. Training departments that sell their courses to other companies may be driven by the need to recover the investment in their training.

Leverage internal capabilities. Another driving force is internal capabilities. Internal capabilities reflect current investments in human and capital assets. A training department must sell products that use these investments. For example, if an organization has invested in high-end computers, desktop publishing software, and a group of talented desktop production people, its training function will be driven to sell desktop printing. If training staff are all skilled trainers, they will promote instructor led courses. If a training function has invested in multimedia production capabilities it will promote training that uses multimedia. If it has invested in a number of specialists in measurement and evaluation, it will sell measurement and evaluation services.

Many training functions are driven by internal capabilities. They have invested in the technology and equipment for developing computer-based, video-based, or multimedia training. They must sell products and services that require these investments, or the investments will go to waste.

A return on investment or profit. These are measures of the financial result of training's efforts. They include percentage of sales, return on assets or equity, cost/benefit ratio, and budgetary control. If this is training's driving force, it will select training products and markets based on return/profit goals. It has become important for many internal training functions that have become profit centers or find themselves required by management to cost justify their operations.

Although the list above can't be called comprehensive, it does provide a description of the kinds of driving forces common to enterprises of all sorts. That a training function should be driven in some way is a new way of thinking for many trainers. But the requirement is real. The training function can become much more effective if it can identify its driving force and conform to the nuances associated with it. If, for example, a training function is driven by technology, efforts to grow outside the technology will be difficult and, perhaps, without merit.

A training function can become much more effective if it can identify the driving force and leverage the opportunities associated with it or conclude that a new, more appropriate, driving force is needed. The guidelines provided here can be used to identify and describe your driving force.

A mission statement contributes to training's ability to provide leadership. It gives others direction. It tells people what business training is in. Following are a few examples. Amoco's Information Technology Training Department developed the following mission statement.

> We will provide superior information technology products and services to mutually beneficial partnerships with customers and other information technology functions.

Following from Amoco's ITT's mission, its vision statement is:

> We will be a premier customer-focused service provider: We will create and continue to improve the standards for excellence by which other customer service organizations are measured. We will:

- Provide support at service levels that meet customer requirements.
- Develop and maintain a partnership and supportive relationship with customers and suppliers.

By having a mission and vision statement, Amoco's ITT Department is in a better position to be driven by its customers instead of its product line or internal capabilities.

INFLUENCE BEHAVIOR AND ATTITUDES

Leadership shows up not only in the mission and vision, but also in how one influences behavior and attitudes. The training function must **influence** what the organization chooses to act on and accomplish. To influence is not the same as to manipulate. Although the actions and results may be similar, the motivation behind the one is *not* the same as for the other. An *influencer* is motivated to move people closer to a vision of what benefits an organization. Leaders typically look for ways to implement mutually beneficial solutions: to tie the people to the group. A *manipulator* is motivated by self-enhancement or self-preservation. So, to say a leader (or leading function) influences others' behavior and beliefs is to say such actions are carried out for the ultimate good of the group.

Peter Senge says that to influence others' behavior and attitudes you send messages through your stories and actions to shape others' expectations.[5] It is through stories that people learn what behavior is expected, tolerated, and taboo, as well as how day-to-day activities are handled. It's not important whether the stories are true. What matters is that people gain insight into what is important (and what is not), what's encouraged (and what's not), and what's prohibited (and what

---

[5] P. M. Senge, *The Fifth Discipline: The Art and Practice of the Learning Organization* (New York: Doubleday, 1990).

is not). Cultures are shaped through the behavior that is modeled and rewarded and the stories that are told.

Leaders both understand the culture and shape the culture. They model the behavior they want to see in others. They reward the behaviors they want. They listen to the stories. They build their own stories. They reward the behaviors that support their vision.

An example of how a training department provided direction is how Allstate Company's Business Insurance Department's executive management identified a set of 18 "imperatives" that, in effect, formed a design for a new corporate culture. Training was chosen to:

- Communicate the 18 imperatives to everyone in the company.
- Consult with underwriting and claims to think through how the imperatives would play out in their work (i.e., how they were going to make the imperatives real).
- Create a self-assessment so people could identify where they lacked the skill and knowledge to apply the imperatives to their work.

Figure 2.3 presents the 18 Imperatives. Since Allstate wanted to position each imperative as equally important, they are listed alphabetically.

FIGURE 2.3. Allstate's 18 Imperatives

• Coaching. • Communications. • Creativity. • Diversity. • Initiative. • Innovation.	• Involvement. • Leadership. • Opinion survey. • Organization. • People development. • Quality.	• Recognition. • Teams. • Technology. • Training education. • Trust. • Vision.

Source: Reprinted with permission of Allstate Insurance Company, Northbrook, Illinois. © Allstate Insurance Company, 1994.

Training took full leadership of the deployment and the associated risks. These included defining each imperative, identifying the skills and knowledge associated with each one, and going public with how they affect people's ability to accomplish good results. Training managed not only the content, but the cultural shifts required to support the new business measures. In other words, it chose to take action and live with the consequences.

As it defined each imperative, training set new performance standards for the company. Figure 2.4 shows the standards set for the leadership imperative. In the figure, level I refers to the knowledge, skills, and accomplishments expected of individual contributors who work independently, members of intact ongoing teams, and leaders of intact ongoing teams. Level II is the knowledge, skills, and accomplishment expected of individual contributors who bring expertise to intact teams and special task forces; team members who work with multiple teams and teams working on complex assignments; and leaders of multiple teams or teams charged with special assignments. Level III is the knowledge, skills, and accomplishments expected of individual contributors, team members, and team leaders charged with high-risk assignments.

This definition of leadership has elevated the level of conversation by managers and professionals across all functions at Allstate Business Insurance about what leadership in the organization means. Also, the definition puts all employees in a better position to assess their capabilities compared to what the company is seeking.

USE OF POWER

The use of power is another way leadership shows up. There is power in the training function. How training chooses to present itself to the organization and how it chooses to perform its function can be incredibly powerful influences. Training often determines the psychological environment for an

FIGURE 2.4. Leadership at Allstate Business Insurance

Leadership. You search for ways to assure that individual and team energy (efforts) is focused on the vision and on activities that add value to the organization. Why you do this is to have everyone contribute positively, think critically, be inspired, be committed, be motivated, and be innovative. How you do this is by sharing your personal vision, soliciting involvement and participation, challenging relevance, surfacing dissension, and recognizing others' contributions.

Knowledge	Skill (Activity)	Accomplishment
Level I		
You know: 1. ABI's vision, values and mission. 2. Customers' current requirements. 3. Your and the team's responsibilities and performance standards. 4. Basic negotiation techniques. 5. How to facilitate groups. 6. Available resources.	As a team member, someone who works with small groups, or one-on-one, you: 1. Explain the team's purpose and direction. 2. Negotiate priorities for team and self. 3. Facilitate involvement and accomplishment. 4. Identify needed resources. 5. Focus on accomplishment.	Your outcomes are: 1. The team goals are met. 2. Time at task is efficient. 3. Everyone participates. 4. The team knows its vision, mission, and priorities.
Level II		
1. Sources of personal power (access to people & resources, alliances, information, competence, external status). 2. Concept of trading favors (information, resources, & access) while maintaining personal and business integrity. 3. Conflict resolution techniques. 4. How to build a strategy. 5. Who has resources. 6. What activities and information add value. 7. Value of delegation.	As a team leader or someone who coordinates the efforts of others outside your normal work arena, you: 1. Build strategic alliances within and outside of team or work group. 2. Seek information that adds value. 3. Negotiate for resources, access, and information. 4. Distinguish when and how activities are in or out of alignment with goals.	1. You influence others or other teams' actions through your alliances. 2. The team has access to needed resources and information. 3. Team members respond positively to your direction. 4. Change efforts are underway with the team.
Level III		
1. External factors affecting team's strategy. 2. Organization's strategies and priorities. 3. Where core competencies lie in the organization. 4. What will be viewed as useful and appropriate by outside agencies.	As someone who leads or supports multiple teams, you: 1. Explain how the team's strategy and goals support the organization. 2. Develop long-range plans. 3. Align resources to accomplish goals. 4. Explain the skills and knowledge mix required of the team.	1. Team's goals and strategies are understood and aligned with the larger organization. 2. The risk is articulated and shared. 3. Your team and you are perceived as role models.

Source: Reprinted with permission of Allstate Insurance Company, Northbrook, Illinois. © Allstate Insurance Company, 1994.

organization. Even trainers who are just instructors and hold no other position in an organization have the power to provide a contextual basis for work in the organization. For instance, a trainer whose attitude conveys to learners that a training activity is irrelevant, unimportant, or will not work, has influenced the learners to disrespect the content, the job, or the organization—and the training function. Therefore, trainers should provide a good example. In addition to the influence wielded in the classroom, training also exercises power when it can participate in executive-level decisions about resources, processes, and technology. If others fear the power held by the training function, they may aggressively confront trainers. Nonetheless, the training function holds power in the organization and should exercise that power with intelligence and vigor.

CONTEND WITH COMPLEX SOCIAL SETTINGS

Training must learn to contend. Leaders take risks—they move into realms where angels fear to tread. They accept the risk of exposing themselves to criticism. The organization can be a complex environment where different parties disagree over priorities, goals, and even what is real. The training function should not be afraid to make its voice heard. How often have trainers been silent when they recognized such an opportunity or when requested to train people to do something irrelevant? Usually the silence is driven by a fear of repercussion, inadequate influencing skills, or lack of information to counter the request. Training demonstrates leadership when it provides whatever intelligence, insights, and concerns it has.

WELCOME EVALUATION

People are always passing judgment. Leaders know this. They are not surprised to learn that others have judged them. However, there is a difference between criticism and evaluation. Formalized evaluation is the basis for continuous improvement. Without evaluation, a trainer does not know where to focus efforts, what works and what doesn't, or what customers think. What leaders do is:

1. *Invite evaluation.* Trainers ask others what they think and how they feel about the function. Is training doing its job? If not, why not?

2. *Base their evaluations on measured results.* Instead of having to guess the value of anything, trainers should base their evaluation on solid measurable evidence.

3. *Make evaluation widespread and continuous.* What trainers learn helps them improve. The more trainers improve the more competitive they are.

DRIVE LEARNING

Effective leaders enable learning. Recent literature focuses on this issue quite extensively. One of the more recent ideas is that of a "learning organization." Peter Senge defined a learning organization as one ". . . where people continually *expand their capacity to create* the results they truly desire, where new and expansive *patterns of thinking are nurtured,* where *collective aspiration is set free,* where people are continually *learning how to learn* together."[6]

Training's business is learning. Training's business is to be an advocate and support continuous learning for everyone in an organization. It is training's job to introduce new knowlege, skills, and insights that will help people achieve the organization's vision. Training drives learning when it:

- Models critical thinking.

---
[6] P. M. Senge, *The Fifth Discipline.*

- Thinks "outside the box."
- Challenges assumptions.
- Separates hearsay from data.
- Helps create a culture that supports learning.
- Sees reality as relative and contextual, rather than universal and absolute.

There are different views of reality. For a trainer what reality is depends on who the trainer is and what the current situation is. It is about the choices a trainer thinks are available. Reality can be thought of as a product of how the world occurs for the trainer. If trainers focus on events and processes, their choices are to design or react. For instance, they look at how work gets done; they might keep score on how many courses were offered or how learning was acquired. If they focus on goods and services, their choices are to provide or consume. They might look at, for example, which courses are offered in the catalog. If they focus on personalities and trends, their choices are to anticipate and predict. For instance, they might look at what others think about a training program, or they might compare what other companies are doing in their training departments. Since everyone perceives reality a little differently than everyone else, each has slightly difference choices. Reality can be looked at as a spectrum; the larger the number of choices, the wider the rainbow is. That reality changes as it is described as colors, or as wavelengths, or as wonder. Effective leaders learn everything the can about their customers' realities and their choices.

STEWARDSHIP

The final way in which leadership shows is in stewardship. Trainers are the stewards of the organization's competencies. Trainers protect and enhance the key capabilities (e.g., skills, knowledge, patents, technology, people, and ideas) that support the organization's vision. When trainers step up to the role of stewardship they:

- Discover the spectrum of capabilities that exist and are needed to move the organization toward its vision.
- Describe the capabilities so that all the stakeholders can understand and value them.
- Plan, develop, and carry out opportunities for employees to learn the capabilities that support the vision.

SUMMARY

Whether trainers accept the role or not, the potential for leadership is always present. If the training function wishes to increase its credibility and status within its organization, it will accept the leadership role. The training function will have to decide—does it want to be out in front or in the rear?

Leadership Standards for Training

Few leadership standards specific to the training function exist. Seeing a need for this, IBSTPI developed a set of standards (see Figure 2.5).

Check Your Understanding

Use the questions on the following checklist with your staff and customers. Check yes or no in the appropriate column. Use the comment column if you aren't sure and to record any remarks you might have. With the answers you can facilitate a better understanding of your leadership role.

FIGURE 2.5. IBSTPI's Standards for Leadership

Rationale	Performances	Criteria
Purpose: To assure that you (the training function) have a vision that allows you to support your customers' business goals. What: You lead the function as a value added business. You align your vision, actions, resources, and measures with your customers' goals. Why: So you provide direction to customers on issues, opportunities, solutions, etc. So you provide products and services that add value.	1. You have a vision statement for the function. 2. You have a strategy for realizing the vision. 3. Public declarations are made about what the function stands for, does, and promises its customers. 4. You have a process to align your goals, priorities, resource allocations, and measures with the vision and customer goals. 5. You lead by example, (i.e., create and set new standards). 6. You evaluate the effectiveness of your vision and strategy. 7. Your vision, strategies, and processes are documented by staff and customers, and validated.	1. Customers consider the function's vision and strategy to be aligned with their business goals. 2. Customers cite the function as a positive model and leader. 3. Staff know the vision and strategy. 4. The processes to align the function's goals, priorities, resource allocations, and measures are used. 5. Documentation describes the processes used to align the function's goals, priorities, resource allocations, and measures. 6. Documentation includes the vision statement, strategy, promises and public declarations, and alignment processes. 7. Documentation includes how and when the effectiveness of the vision and strategy was and continues to be evaluated. 8. What is learned is documented and shared within the function.

Source: Reprinted with permission of the International Board of Standards for Training, Performance, and Instruction (Barrington, Illinois). © IBSTPI, 1994.

Training and Leadership

Question	Yes	No	Comment
1. Does the training function have a mission?			
2. Does the training function have a vision statement?			
3. Is there a published statement of values for the training function?			
4. Does training invite evaluation, feedback?			
5. Has the training function developed strategic alliances within and external to the organization?			
6. Do the stories training tells support what it says it values and the culture it wants?			
7. Have you developed a definition of leadership for training?			
8. Is training willing to take on a leadership role?			

What To Do Next

1. Create an operational definition of leadership for the training function. Use Allstate's definition as a model format.

2. Asssess yourself against the IBSTPI leadership standards.

3. Develop a statement of values similar to the one used by McDonald's purchasing officers.

4. Develop a list of criteria to judge whether you are being effective as a leader or perceived as a leader by your customers.

5. Get the mission and vision statements of your key customers and the customers you want to serve. Use that to identify services you could offer to your customers that they would perceive are of value.
6. Ask your customers what business they think you are in.
7. Ask your customers what training and performance services and products they buy or would like to buy.
8. Check their list against the products and services you offer.
9. Discuss stewardship at a staff meeting.

3
STRATEGIC PLANNING: OPERATING TRAINING AS A BUSINESS

> *In the 1960s and 1970s, Chicago had a problem very common with other large cities: pigeons. They were everywhere. They ate anything. They dirtied the sidewalks of all parts of the city. What was to be done? At the same time, environmentalists and animal rights activists were concerned about the plight of the peregrine falcon. Again, what was to be done? The solution required both groups to cease thinking in the usual ways and begin to consider their problems from a new perspective. In essence they had to develop a new strategy: natural predators for Chicago and a new, more friendly, environment for the endangered birds.*

Although it isn't apparent, the situation for the training function is similar. On one hand, the quality movement, particularly the Baldrige Award and ISO standards, mandate an important role for training. On the other, most training functions have found their usual way of doing business is not working very well.

The quality movement has presented the training function with a wonderful opportunity. You can take advantage of the emphasis both Baldrige and ISO have given training. You can begin to operate like a business or, for most of us, a business within a business.

Every training organization wants customers to ask it for services and products. However, customers are often driven to choose another vendor because they perceive it offers better value or price. Together, value and price are measures of business success. That sounds simple, but the challenge is to find out what customers value and what they consider to be a fair price. Successful training departments discover what their real costs are and offer customers cost competitive services. They demonstrate the value of their services. Therefore, Chapter 3 addresses:

- The nature and extent of the competitive challenge.
- How to create a strategic plan for the training department.
- How you can test your own readiness and ability to effectively compete for your customer's business.

THE CHALLENGE

Why should a training function be a business? Because if training departments fail to provide value at a price that is competitive, they will perish. Three factors hurt the business opportunities of internal training departments. First, price historically has not been an issue. Most training organizations are considered support functions. Their managers do not know what their organization's costs are. They may not know how to price their products, let alone set a competitive price. Second, customers have increased access to alternative sources for training. There is increased direct competition from independent trainers, universities, consulting firms, and training departments in other organizations. Third, many customers are dissatisfied with the quality and price of the services they get from internal training sources.

Internal Discontent

An important factor is customers' dissatisfaction with what training provides. There have been too many situations where the training department has been content to be reactive and simply wait for customer requests. The result is customers go elsewhere. Whole departments have been eliminated because they lost their mandate.

For example, the internal sales support manager at the Jerry Co. wanted a five-day course on how to use a $4 million order entry system. Training provided a five-day orientation and training session. The learners felt the training was worthwhile. While the training was occurring, management was dismayed to hear that sales were down significantly. With only one person left to answer the phones, customer requests were put on hold. The sales force was very unhappy over losing commissions. The company was upset about lost sales and a slip in their market share. The learners thought the training worked well, but thought one day in the classroom would have been enough if they had on the job support.

The moral: Being competitive is not about merely being obedient to the customer but adding value. Did the training department deliver exactly what the customer's said it wanted, a five-day training event? Yes. Did the training do what it promised? Yes. Did training deliver a competitive training event? No—it cost too much just in terms of lost productivity. Will Jerry's training department be viewed as good business, will it be entrusted with other high-risk training ventures? Not likely!

Valuing Competition

Competition can actually improve quality. Without competition, customers have no choice. As soon as a competitor arrives, customers can compare offerings and choose the one they like best. The criteria for judging quality are supplied by the customers not the providers.

However, many professionals are troubled by competition. They don't understand what it is or how it operates. Here's a rule of thumb: when a training department finds itself trying to compete with other internal training functions, it probably has taken a wrong turn somewhere. To identify competition, training management should:

1. List products and services the training department produces.
2. Identify current and potential customers.
3. Survey current and potential customers; find out what training products and services they are buying, from whom, and why.
4. For products and services customers buy outside, analyze the reasons:
   — The training department doesn't offer it (and has good reasons not to or doesn't have good reasons not to).
   — Customers don't know the department offers it.
   — Customers know it is offered, but don't like it because of price, responsiveness, or timeliness.
5. Use the findings to set a viable strategy.

In summary, then, to be successful in business, training has to know who its customers are, what they want, what they are buying, from whom, and what value they are getting from their training suppliers. It can use that information to compare products and services to what the customers are buying, decide if it wants to be in that business, and if so, how it can compete.

STRATEGIC PLANNING: QUALITY'S CONTRIBUTION TO COMPETITIVENESS

A Baldrige or ISO committee looks at a candidate's plan to achieve quality. They look to see if the plan tells who the candidate is, where it is going, how it intends to get there, and how it will know if it has succeeded. They look at how the candidate intends to incorporate the voice of its customers both in defining what it does and how it will measure its success. They want to know how it will deploy quality throughout the training function. The candidate's plan becomes the basis against which its actions and results are judged. This strategic plan tells customers and employees what the candidate is doing to become and remain a viable entity.

Training functions usually create project plans and budgets. However, strategic planning is new. It requires training to declare what it stands for, why what training does is important, and how it is going to honor its public promise for improving performance. To do strategic planning well, training has to know what customers want and value. It can use that information to plan how it will deliver what they want at a cost they believe is reasonable.

Who Are Training's Customers? It's critical to identify customers. It's easier to become competitive when training is clear about its customer base. Does it serve different business units, some, or only one? Does it serve different line functions in a business unit, some, or one? Does it train across all levels, or some, or only one. For example:

- Household International's Business Institute supports only management development across all business units.
- McDonald's Training Development department trains only members of the restaurant crew and management team.
- The Walgreen Company Training and Development supports only the training needs of their drug stores.

Walgreen's Story

The training department's internal customer is the drug store. Training began to conduct cost benefit analyses in the early 1980s (when the company was rapidly expanding and there was no mandate for cost analysis). The four-step process put into practice in 1980 still works well:

1. Training and development (T&D) always agrees to provide whatever the drug store business wants. The request is not qualified in any way in the beginning.

2. T&D always asks customers to specify what they expect to improve if the training is successful. The specifications are quantitative. For example, if the drug stores want training in just-in-time distribution, they would specify how many pieces of what stock do they want replenished within what cycle time? If stores want to improve absenteeism, they would specify what level of absenteeism will be satisfactory to management after training intervenes.

3. T&D develops the intervention and pilot tests the intervention in real situations (i.e., no simulations). The pilot test is well documented and professionally administered.

4. T&D measures changes to the specified measures over time (usually six months) to discover how much of what kind of change can be attributed to the training. If the change meets customer requirements, the program is deployed. If not, it is dropped.

T&D can always demonstrate results, which assures their customers they are competitive. When Walgreen uses an external vendor, it uses the same procedure to measure the vendor's success whenever it can. No surprise, Walgreen's T&D is well accepted, even admired, by their customers and by other departments.

Some companies handle the problem of identifying customers by creating multiple training departments. A large international petroleum company discovered it had more than 40 different training departments. Each was originally created to service a specific customer. However, many of those customers had the same needs (e.g., new hire orientation, basic supervision, computer application, diversity training). The result was duplication and wasted resources. Internal training departments found they competed against each other. Worse still, customers were confused over who did what.

Ford's Corporate Quality

Ford Motor Company has been a leader in the quality movement in this country for years. Its basic advertising promotes "Quality is Job One." Ford's success is evidenced by the upturn in profits for the company and the increasing popularity of their products. But, were things as good as they could be? In particular, were Ford employees getting the kind of training they needed as far as quality is concerned? Don Smith and his people in Corporate Quality were asked to find out. In response they conducted a study called "Ford's Quality Education and Training Study," in October of 1992. They surveyed all Ford employees (except those in their home mortgage affiliate)—192,895 workers. The results:

- Ford employees attended a total of 964 different quality-related education and training courses.
- 170 of the courses came from external sources.
  — 34 organizations within Ford.
  — 61 outside consultants and training companies.
  — 32 companies teaching their own proprietary information.
  — 30 colleges and universities.
  — 13 associations and professional organizations.
- Approximately $3,600,000 was spent on external sources.
- There were major redundancies in 62 of the programs.
- 52% of the activity areas stated their present quality-related education was not meeting their needs.

It is not yet known what actions Ford will take because of this information. Probably, there will be further investigation to get more specific information about particular courses and how some might be better suited to Ford's requirements than others. Whatever the outcomes, it's apparent this information is very valuable. It provides management with a much better position from which to make its next decision. Which leads to the crux of this chapter. Why measure? Why spend the time, effort, and money needed to get valid reliable data about these things? In essence the answer lies in the quest for quality and the role training plays in that quest.

IDENTIFYING CUSTOMER'S NEEDS

What Training Services Are Customers Buying? Training must also understand the pressures its customers face. It is these pressures that drive training needs. Pressures include:

- Generating higher market share.
- Increasing productivity.
- Satisfying regulatory requirements.
- Decreasing accident rates.
- Retaining customers and increase customer satisfaction.
- Qualifying or certifying workers for specific job categories or tasks.
- Being environmentally responsible.

Training must compare its services with those its customers are buying. Training must find out how much they are spending with what vendors and what kind of value they are getting for their money. Training staff must ask:

- What services should we provide to meet customers' need?
- What would it cost us to deliver them? Are we even close to being cost-competitive?

- What quality can we provide? Can we add value other vendors cannot?

For instance, one company wanted to train its own training staff to do needs assessments. It compared what it would cost to develop and deliver the training itself with using an outside consultant who specialized in the area. It was cheaper to hire the consulting firm. But the important message is, they knew what it would cost them to develop it themselves. They had a basis for comparison.

Unfortunately, many customers think of an external source before they think of their internal trainers because they:

- Are dissatisfied with poor products and services (i.e., quality, price, or both).
- Don't think the internal department has the capacity or capability.
- Overvalue outside sources.

Often, even when customers are satisfied with products and services, and feel the internal department has the capacity, sometimes price becomes the issue. Many training departments cannot operate as cheaply as external vendors. Their overhead and salaries are high and their charge-backs insufficient. What can they do? They can:

- Improve their processes to reduce costs while improving quality.
- Find a market niche they can fill competently.
- Eliminate costly services that don't contribute directly to satisfying customer wants.
- Broker services offered by others.

Getting to the point where internal training can provide customers with what they want can seem like an overwhelming task. The way to do this is by creating a strategic plan for the training function.

DEVELOPING A TRAINING STRATEGY

Development of a strategy for the training function should be carefully thought out and crafted. A systematic process for doing has four steps.[1] They are:

1. Write a mission statement that conforms to the mission of the parent organization.

2. Develop a vision statement to project the effect of the mission on the future.

3. Select a strategy or strategies that effectively and efficiently produce the desired results.

4. Plan actions to implement the strategy.

Each of these steps is examined in one of the following sections.

The Mission Statement

In chapter 2, there was mention of a mission statement. When an organization starts its mission, it already knows who its customers are, what they want, what they are buying, and from whom. The answers to the following five questions are the basis of a mission statement. (Some of these questions were looked at earlier in the chapter.) The more specific your answers, the better the mission statement.

---

[2] J. Hale, *Strategic Planning Worksheets and Guidelines* (Western Springs IL: Hale Associates, 1994).

1. Who are the customers?
   — Who is buying training from us currently?
   — Who do we anticipate buying from us in the future?

2. What are the customer requirements for the products and services provided?
   — Current.
   — Trends.

3. Who are the learners?
   — Who is using current products and services?
   — Who is anticipated to use training's products and services in the future?

4. Why do customers buy from training?
   — How are current customers retained?
   — How could training capture and retain future customers?

5. How can training maintain professional competence so it can deliver competitive training (quality and price meet customer requirements)?
   — How does training maintain and grow proficiency in current competencies?
   — How could training establish and develop additional competencies?

With answers having sufficient detail, training is ready to write its mission statement. It could read something like this:

> Training and development at XYZ is dedicated to supporting our operations staff. We provide the learning events people need to produce top quality products. We secure, develop, or deliver the sales training events needed to maintain a growth rate of at least 12% per annum in sales. We provide our services at cost-competitive rates and achieve agreed-upon returns on investment to customers. We continue to develop our staff, provide them with the best knowledge, skills, tools and equipment available to instructional technologists.

This is a generic model. Each training department will want to add specifics reflecting its company's culture and mission.

Vision Statement

The concept of a vision statement was also discussed in Chapter 2. A vision statement is built on the mission, but describes the future. The format is similar to the mission, except it describes the future state of the organization. The statement says in essence, "We expect to become (...) by (...)." Training's outlook toward the future has three viable alternatives.

1. Develop more expertise, responsibility, and clout.
2. Stay the same.
3. Narrow your scope.

Most people think the third possibility is not something to work toward. But if a training department currently operates with a stagnant bureaucracy, is supporting equipment and people that don't add value, or for some other reason can't deliver what customers require, smaller may be better (at least for the near future). "Lean and mean" is the vision for many functions, including training, in today's successful organization. Developing a good vision requires answers to three questions:

1. Who do we *want* as customers?
2. What products and services do we *want* to offer?
3. How do we *want* to differentiate ourselves?

The answers to the first two questions are fairly easy. Chances are the customers wanted for the future are current customers, with perhaps some modifications. For example, a training department may want to do more or less training, thereby enlarging or narrowing your customer base. Likewise, products will probably be similar but they will better match customer requirements for quality. The product mix might shift, for example, away from delivery and toward needs assessment, or training might want to stop doing production altogether.

As a training department develops reasonable answers to the third question, "How do we want to differentiate ourselves?", it will probably want to modify its answers to the first two. The answers to these questions depend a great deal on circumstances and setting. A training department at a heavy equipment manufacturer, for example, decided to become best at providing job aids for the manufacturing process. Another company, whose mandate was to develop the organization's sales force, saw its primary goal as bringing about basic culture change.

Some training organizations have chosen to broker training services for their internal customers rather than providing the services themselves. More training departments are devoting their resources to developing a sound request for proposal (RFP) process to bring in outside training vendors. That way they maintain control of training standards, gain the expertise or help they need, and avoid the long-term liabilities of permanent staff. They find their competitiveness enhanced by using their resources to define work, manage the bidding, and administer contracts with outside firms.

Here are a few avenues a training department may want to explore to differentiate its function:

- *Products.* Training could consider performance, features, reliability, conformance to standards, durability, aesthetics, perceived quality, serviceability, and other attributes as well as type, style, and configuration. For example, Walgreen's T&D must produce quantifiable results (i.e., improve the business performance of the drug store).

- *Service.* Who will provide basic services, internal professionals or external consultants? Will you do needs assessments, job analysis, and other performance analysis tasks or stick to delivery of training? Will training department employees attain certification of some sort? For example, because of cost cutting measures (i.e., downsizing), one training department is outsourcing its instructional design and development services. Other training departments are outsourcing their delivery and using contract trainers.

- *Cost.* How will products and services stack up against the competition? How will training justify its budget? How will it charge customers? For example, one training department tracks prices for the services it offers through professional organizations and informal networks among training directors.

- *Availability.* How will training's presence be known to customers? How responsive will it be? Just in time? For example, since Walgreen's training and development only offers services to drug store staff, the accounting department has to buy its training from outside vendors.

- *Ongoing support.* How will training follow up on services? How can training maintain the quality of products already in use? What will be the style and nature of summative evaluation and what will training do with it when the results are in?

- *Image.* What does training want its customers and others, both internal and external, to think about its operation? How will the department promote itself? What sort of picture will it present to its customers and to colleagues outside the organization? For example, FedEx's training department publicly promotes its award-winning supervisory development program called "LEAP."

A vision statement should be shaped by the driving business forces of the organization. Otherwise, the entire exercise can be a waste of time, effort, and resources.

Following is an example of a vision statement:

By the year 2000 our department will be delivering 80% of the training for XYZ. At least 70% of that training will be on-line, just-in-time, using computer-based training technology. Our courses will qualify XYZ as a viable competitor for the Quality Award and all tenured training personnel as well as the department itself will have attained professional certification. The department will be a fully independent internal operation using charge-back techniques to operate as a profit center.

Selecting a Strategy

The mission statement documents the present. The vision statement charts the future. To make both real, a training department must have a *strategy* that outlines its course of action. A strategy is the general scheme to achieve the vision. It documents how to get to where the department is headed. For example:

- An international restaurant chain's goal is to increase profits. One of their strategies is to increase breakfast business.
- A credit card company's goal is to increase its market share. Its strategy is to offer private label cards that bear a retailer's name.

Strategies are broader than tactics. Tactics are specific actions used to accomplish particular objectives. The strategy is what makes tactics viable. The training function as an entity must develop a strategy. Tactics can then be used to make that strategy work.

A strategy is usually a long-term action plan of three to five key initiatives that leverage opportunities and counter threats. *Opportunities* (along with strengths) are the factors that will increase the probability of achieving a mission and, eventually, a vision. Opportunities, once identified, must be acted on or they lose potential value. *Threats* are factors that will limit or prevent achievement of the mission and vision (if they are not countered, neutralized, rebutted, or overturned). Threats are internal weaknesses or external strengths that can't be countered. No matter if they seem huge or minor, threats must be identified and addressed. For example, one international retailer chain developed a training strategy to invest in and exploit electronic training delivery. Another company, a credit card company, developed a training strategy to provide world class training to their customer service department.

Any training function can also develop a viable strategy. The process recommended here for developing a viable strategy has nine steps (see Figure 3.1)

An example of a training strategy is provided by the Information Technology Training (ITT) group of Amoco:

> The ITT team identified the pressures within their business environment, such as:
>
> - Less money.
> - Just-in-time (being customer focused).
> - ITT perceived as top heavy and expensive.
> - The increasing expectation of management to justify its cost.
> - Business Unit approach (decentralization).
> - Outsourcing.
>
> Amoco, the parent company, had identified five factors that influence performance:
>
> 1. Strategic Direction.
> 2. Structure.
> 3. Processes.
> 4. Rewards.
> 5. People.

FIGURE 3.1. Developing a Strategy

Step	Comments
1. Develop two comprehensive lists of: • Opportunities and strengths. • Threats and weaknesses.	• Include everything anyone can think of (even though some ideas may seem far fetched or minor). • Document every opportunity and threat. Strengths and weaknesses include the staff's competence and reputation, facilities, and equipment.
2. Summarize key opportunities and threats. Examine the two lists and group the items listed in priority order. When complete, there will be two sets of tasks: • Opportunities to leverage. • Threats to counter.	• What training opportunities could yield the best return on investment? • What threats could be most damaging to accomplishing your goals? • What will have the greatest long-term impact on your department?

(continued)

The ITT team decided to use these five factors as part of their strategy. They developed vision statements for each of the five performance influences and working definitions of each key point.

Strategic Direction. We will build relationships with our customers so we become part of their value chain and help them do their jobs better. We will achieve this by supporting and rewarding entrepreneurship and continuous improvement. Our relationships will be meaningful and long term. Customers are the people who support us with their budget. Value chain means the key processes our customers require to serve their customers. Entrepreneurship:

• Conveys ownership.

FIGURE 3.1. Continued

Step	Comments
3. Evaluate the current strategy.	• Does it require more or different competence or resources? • Does it fail to exploit current opportunities? • Does it counter important threats? • Is it congruent with your values? • Is it still valid? • Does it exploit your uniqueness? • Is it sensitive to recent or anticipated technologies? If so, pursue this strategy. If not, continue with step 4.
4. Identify potential strategies.	• What actions, initiatives, and focus of resources will take advantage of opportunities and strengths? • What will most likely counter the threats and weaknesses?

- Conveys being the best and delivering the best.
- Is an attitude.
- Is inspirational.
- Means growth.
- Means hard work.

Being the best means being:

- Focused.
- Flexible (multiskilled).
- Responsive (just-in-time).
- Willing to continually examine management's and supervisor's roles.
- Able to deliver the right products and services.
- Cost competitive.

The measures we will use to help us determine if we are the best are:

- Customer satisfaction.
- Reliability.
- Availability.
- Responsiveness.

Structure. We will be organized in ways that support building relationships so we can respond to our customers' evolving needs while optimizing our resources and promoting employee satisfaction. We will achieve this through structures that:

- Accommodate change.
- Leverage our core competencies.
- Give us a competitive advantage.

The group defined output as:

- Billable hours.
- Customer contact and relationships.
- Products and services customers value, use, and pay for.
- Units of work.

Business Processes. Our business processes are customer-focused, timely, cost effective, and measurable. We will achieve this by:

- Documenting them.

- Continuously improving them.
- Streamlining them.

Our key business processes are:

- Marketing.
- Production.
- Distribution—how we get our products and services to our customers.
- Planning and Resource Allocation.

To continuously improve processes means we will examine, reengineer, or eliminate them. Streamlined means our processes are direct, uncluttered, uncomplicated, well defined, aligned with business goals, and not redundant. Our processes will be invisible to our customers.

Rewards. Amoco had identified eight specific measures departments are required to look at when examining their processes. ITT decided to focus initially on three:

- Customer service/satisfaction.
- Cycle time.
- What is valued by the customer.

We will reward entrepreneurship and continuous improvement. We will achieve this by:

- Having a well-managed and understood process to recognize, reward, and celebrate individual and team accomplishments.
- Management and staff mutually identifying innovative symbols of accomplishment.

Rewards include:

- Compensation.
- Rewards and recognition.
- Celebrations.

Rewards can be defined by management and employees. Rewards will be based on making Amoco successful.

People. People give us our competitive advantage. We will achieve this by:

- Executing our people processes effectively.
- Giving our people the right tools.
- Providing an environment that promotes:
  —Open communication.
  —Professional development.
  —Promotion.
  —Reward and recognition.

Once a training department has developed a strategy that supports both the organization's and its own function's mission and vision, it is ready to develop a business plan.

Developing the Business Plan for the Training Function

The training function is a business. It should have a business plan. In fact, most probably already have one. It may not be well articulated, but training management has made some decisions and have ideas about how to carry them through. If training formalizes that plan it will be easier to implement. When plans are documented they can be shared, managed, and improved. When plans remain informal it is difficult to evaluate their effectiveness.

It's not difficult to build a good business plan. It's even easier to modify a good one once it is in place and operating. Planning should become a tactic used not only to keep up with change but to anticipate and take advantage as it is occurring. This will strengthen training's leadership role within an organization.

A *business plan* is a written document prepared to convince others the function's plans are viable and well thought-out. Training's management has to be clear about who its intended readers are and what it wants from those readers because the business plan presents an argument for additional or continued support.

A business plan tells its reader what an organization's business goals are and how those goals are to be accomplished. It presents the rationale for why the objectives are right for the organization and have value to the reader. A basic business plan is the following:

- Specifies goals and states the case for why its important to achieve them.
- Documents what resources can be used to achieve the goals, and how their use will impact the reader and the organization.
- Provides a (tentative) plan of action: what will be done, by whom, for what reasons.

The business plan is based on an understanding of the organization's driving force and shows how the organization plans to use its mission, vision, and strategies to develop the business.

Elements of a Business Plan. A business plan usually has ten sections. (The sections are listed in an order designed for the readers, not necessarily the order in which they were written.) Following is a sample business plan:

1. Executive summary.
2. Parameters.
3. Who the training organization is today.
4. The marketing plan.
5. How the organization will design, develop, and deliver its training.
6. How training will operate.
7. Training's management team.

8. Training's schedule.
9. What are the risks.
10. The budget.

The following sections take a more detailed look at each element.

Executive summary. A one to two page summary that highlights the important features of the plan and allows the reader to quickly determine if the plan is of interest. It has four sections (about a paragraph each):

- Market opportunities.
- Products and services.
- Financial projections.
- Proposed financing.

Parameters. This description of the department and its products includes enough information to convince readers that training's management knows what business training wants to be in and knows how to make it happen, such as:

- Events that could affect training positively or negatively.
- Mission and vision statements.
- Strategy for accomplishing the mission.
- The driving force as it applies to the training function's situation.
- How the department is organized and located.
- How training differentiates itself today and will in the future.

- Training's products and services and what makes them unique.
- Opportunities and threats.

Who the training organization is today. This is a description of training's customers. (The training function is, in a sense, a reflection of its customers.) The goal is to convince readers training can address its customers' needs. Include:

- Who the learners are, how many of them, and how much training will be delivered.
- What is driving customers to training (e.g., regulatory requirements, ISO certification, industry or functional experience, price, or quality).
- What customers have budgeted for training.
- Who customers buy from now and why.

Marketing plan. This plan gives enough information to convince readers training management knows how to price, sell, and deliver training. It includes what will be done, how it will be done, and who will do it, and the following information:

- The customers training wants to reach and how. The plan says what products and services will be stressed and how customer satisfaction will be measured. It specifies the role customers will play in setting performance criteria for the training department and its products and services.
- The plan describes how training will be priced (i.e, chargeback). It explains how the pricing compares with market rates and what costs it covers.
- The plan describes how training will reach desired customers.

How the training function will design, develop, and deliver training. The plan gives enough information to convince readers training management knows what it will take in time and resources to create new products and services that customers want. It describes how any current products and services will be changed. The plan explains such difficulties as scheduling, costs, time constraints, and how they will be addressed. It also describes plans for new products and services.

How training will operate. The plan explains how training is to operate and gives information about what staff, space, facility, equipment, and administrative support are needed to operate your business. This section explains the function's operations base and its advantages and disadvantages. It describes how training will staff the effort (e.g., contract trainers, outsource production, or train customers to administer training). Also included is a description of facilities and improvements, what is on hand and what is needed (if anything) to better support the business. This section also explains training's management and tells them how training influences decisions about what to offer and what customers to service.

The management team. This section gives enough information to convince the reader training has the people who can turn ideas into reality; they have the appropriate balance of competencies. It includes a description about the key management roles, the qualification (expertise and experience) of training's staff, and the governing body. It tells how the function gets its philosophy and overall direction. For example, is there a training council made up of representatives from key customer groups?

Training's schedule. The action plan communicates not only what you intend to accomplish, but also how, and when, the organization can expect to benefit.

What are the risks? This identifies the risks and explains how to manage these risks. This includes:

- Challenges customers are facing.
- Situations that might cause problems, such as not meeting a schedule.
- Dependence on a supplier or training vendor.

Training's budget. This explains training's financial planning and thoroughly lists salaries, materials, consulting fees, and other costs.

Members of the training function should review the elements of the business plan and consider the implications of each one. How will it impress the reader? How does training management want the reader to react to the information? In each case the critical aspects and impacts will be the same: How does the element in question help the reader become more convinced that what training is doing will lead to success? Training is, in effect, selling its business to the reader as a start-up effort might sell its ideas to a venture capitalist.

Training management should ask the people in the function to study the business plan and what it implies for them. When they combine it with their understandings of the mission, strategy, and driving force for the function, they should be able to place themselves productively in the matrix. They should be able to easily see how each person can contribute personally to the effort and why the function will be able to accomplish its goals.

Training management should also ask peers in other functions within the organization to read the business plan. They are training's clients and their understanding of how training intends to operate will prompt them to ratify training's aspirations for serving them. Having done that, they will be likely to call on training whenever they have a requirement training can service.

In effect, a good business plan is a very positive self-fulfilling prophesy. A business plan will make operations smoother and easier to sell, both internally and externally. It's something that, once done, causes managers to wonder why they hadn't done it in the past. It is the key to the manager's ability to build a business. Done well, a good business plan will put internal training in position to do what it is designed to do despite the best efforts of other vendors.

SUMMARY

Training can no longer escape the demands being placed on its customers to provide value-added and cost-competitive services. It, too, must define its role and market, and must differentiate itself from other providers. The guidelines for developing a strategy and business plan parallel what other functions do to be competitive. The process of developing a business plan will prepare training management to manage better its function and relate to its customers.

Check Your Understanding

How Competitive Are You? Hopefully, as you've read this chapter, you have developed insights about challenges to the training function, who else is in the market, and the keys to developing an edge. The questionnaire below is a way for you to summarize these ideas and judge the adequacy of your current business capabilities.

If you're sure you can answer yes without reservation, check the yes box. If you have reservations check yes and add a comment or two. If you must check no, be sure to explain your response in the comments section. If you don't know the answer, don't check either box, but do say why you've refrained. When you've completed this exercise, you will have the basis for development of a viable training business. You can begin to improve your situation at once. Simply begin efforts to change any area where you've written comments.

Question	Yes	No	Comment
1. Do you know how many suppliers of training and development services your organization contracts with annually?			
2. Do you know how many departments within your organization provide training products and services?			
3. Do you know what distinguishes you from other suppliers both within and outside of your organization?			
4. Do you know how what you charge compares to other suppliers?			
5. Do you know how your customers rate your services compared to others?			
6. Do you know the strategy your organization intends to use to secure its current vision?			
7. Do you know what executive management feels is the driving force for your organization?			
8. Do you know your clients' most pressing needs?			
9. Do you know how many of your own people consider the training function to be a business within a business?			
10. Do you know how the products and services you offer compare to what your customers are buying?			

What To Do Next

Use your answers to the questions as a guide for discussions with other members of the training function. The following steps are recommended:

1. Assign a team to develop a business plan for the function.
2. Ask your customers to share their business plans.
3. Ask other training managers in your area to share their business plans.
4. Do a market study to determine how other training functions charge for their services.
5. Do a customer satisfaction survey.

4
OPTIMIZING TRAINING PROCESSES

A male mallard resplendent in its iridescent greens and whites and shades of grey and brown, surges across a wilderness pond and rises into the air. Above, a silent watcher glides on the warmth of a thermal air current. As the duck clears the marsh and turns toward a distant feeding ground the other also turns.

The duck thunders along, racing to join others in formation. He doesn't notice his doom until it's too late. In an instant the deed is done. The duck stalls and pitches toward the ground. His assailant grasps him, pulls him close, and heads towards the aerie where hungry mouths await.

In another part of the world, five pigeons waddle around a small concrete plaza in a park near downtown Chicago. One wanders toward the center of the space pecking at bugs and crumbs. It flutters its wings and hops forward. A blur of dark brown and beige flashes down and knocks the pigeon tumbling across the pavement and kills it dead in an instant. The falcon settles to the ground and hops toward its meal as the four other pigeons scatter squawking into low escape flights. After a few tentative stabs at the victim, the falcon grasps it with its talons and leaps into the air to fly heavily back to the tall building where its chicks wait.

A peregrine falcon takes its prey with a sequence of actions learned through many generations of trial and error. The process it uses is efficient and effective. Unlike its cousin, the golden eagle, it seldom misses, seldom merely wounds. It is a hunting machine, perfected through generations of evolution and months of practice.

The peregrine falcon perfection demonstrates what can and should be done with the processes trainers use to deliver developmental interventions. A peregrine accomplishes what is necessary with a minimum expenditure of effort. It does it quickly, accurately, and thoroughly.

We aren't advocating that trainers should become aggressive like falcons. We aren't hunters in that sense at all. How peregrine's hunt, however, serves as a very good example of how procedures and processes can be made better. It illustrates how a small effort can reap huge benefits.

The purpose of this chapter is to discuss processes used by a typical training function to accomplish its tasks. Before we can do that we must discuss what we mean by the term *process*. Process signifies the actions (also called procedures), decisions, and exchanges of information that occur to transform information into training products and services that customers can use. The actions, decisions, and exchanges occur over time, whether they are done sequentially or in parallel. Processes take time and use resources. Baldrige and ISO standards require organizations to describe, document, and evaluate their processes. The emphasis is on proving that processes do not waste resources and result in products and services customers want and value. Training is no exception.

It's critical for a successful, quality-oriented department to understand and document its processes. For unless a process is documented and described it cannot be evaluated. It can be neither managed nor improved.

There are many processes used by training depending on the scope of services offered. For example, a typical training department can conduct:

- Needs assessments.
- Task analyses.
- Instructional design.
- Instructional development.
- Instructional production.
- Instructional delivery.
- Measurement and evaluation.

Optimizing Training Processes

- Administration (e.g., of student records, registration, scheduling of instructors/rooms).

Each of these processes use information about the customer. Training takes that information and through a series of actions and decisions produces recommendations and interventions. When they work well, processes contribute significantly to the success of an organization, especially training's customers. Unfortunately, in too many cases, they could be better designed and better used to generate quality outcomes. This chapter is devoted to helping a training organization examine its own processes with the aim of improving them and establishing them as profitable operations. This chapter begins with a look at what processes are.

PROCESSES

A process is a series of activities and decisions used to transform (change the state of) materials or data for some purpose. Every process uses the output of a previous process as its inputs. Then, every process produces an output that becomes input for other processes. The Rummler and Brache Model in

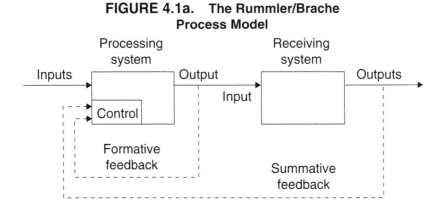

FIGURE 4.1a. The Rummler/Brache Process Model

Figure 4.1a provides an elegant picture of a typical process and its components.[1]

A model used by Rummler/Brache shows the important relationships between the training process and its clients and customers. Before we look at the model as it applies to the training function we should examine the generic application because it's appropriate for many circumstances.

In Figure 4.1, the receiving system is the customer of the processing system (internal, usually). So, summative feedback originates with the customer. Formative feedback comes from sources internal to the process such as the people doing the task or engineering specifications. It is not considered to have come from the customer. A system that relies solely on formative feedback is lucky to satisfy its customers fully. However, formative feedback is important, otherwise internal control is very difficult. In this day of quality consciousness, summative feedback is just as important. Both are considered in a successful operation.

It is important that this diagram should be looked upon as an integrated whole. Although any part of a process may be examined independently, it cannot be removed or changed without affecting the entire process. There are always inputs

---

[1] G. A. Rummler and A. P. Brache, *Improving Performance: How to Manage the White Space on the Organization Chart* (San Francisco: Jossey-Bass, 1990).

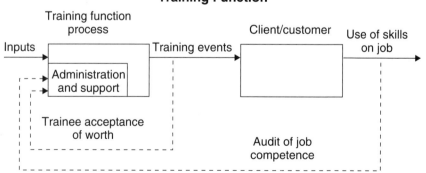

FIGURE 4.1b. Feedback for the Training Function

and outputs. Both kinds of feedback happen whether they are considered or not. A control function is always imbedded in the process some way.

By simply changing the labels for the various parts of the model we can see how it applies to evaluation of the training function (see Figure 4.1b). To control the efficiency, effectiveness, and quality of the process the training manager must both administer the process and support it with resources and policy. The control function is extremely important for training because the function is so dynamic: every training event is unique. Each is used with new or different clients and/or customers. Content changes. Instructors change. The facility and other environmental elements change. The very reasons for using a training event change depends on the wishes of the clients and the needs of the customers.

Now consider summative feedback. Baldrige and ISO quality criteria are applicable in the summative feedback loop: the audit function. Training functions that have existed without audit trails and formal performance audits in the past will have to implement this function or they will probably fail in their attempt to fully support the quality requirements. Without the audit they can't evaluate their compliance with quality criteria and standards.

Also important, the elements are in every process large or small, general or specific, formal or informal. If one aspect is modified, it will modify the others. It's important, therefore, to consider the ramifications of any change.

When processes are viewed as a whole, a trainer can take advantage of Margaret Wheatley's advice to develop a self-organizing (and, therefore, self-renewing) system. ". . . They are self-referenced, stable over time, and able to change the environment. We see systems that evolve to greater independence and resiliency because they are free to adapt and they maintain a coherent identity throughout their history. . . ."[2] To realize this perspective, trainers must examine

---

[2] M. Wheatley, *Leadership and the New Science* (San Francisco: Berret-Koehler, 1992), pp. 94–95.

their processes with the purpose of making them work as efficiently and effectively as possible. But why is such a perspective important? The world has changed. The quality movement is leading that change. An important aspect of the quality movement is its reliance on customers' wants and needs. As much as anything, customers want dependable service. They want to know what the standards are, how closely current products and services conform to those standards, and what is being done to make improvements.

Unfortunately, many training functions have not bothered to perfect and document their processes so it is very difficult to maintain an acceptable level of service. They cannot demonstrate that their services meet standards. (Given the lack of documentation many of them can't rigorously describe their services or standards.) One project may develop very well and produce excellent results while another project, developed by the same unit, may encounter a myriad of problems and literally fail. Without clearly defined and documented processes, success is hit or miss.

In the past, training functions could survive these kinds of disparities by ignoring them or patiently waiting until negative reactions died down. Training's support came from internal management so the reaction was to calm the boss and hold the line. Now, that has changed. A customer can, and will, insist upon competence from training. Internal training can't finesse its way through difficulties any more. Why? Dissatisfied customers go looking for another supplier. They do not have to accept its efforts as all that is available anymore.

For instance, one major company encountered a competitive crunch that caused management to restructure. They decided to eliminate 12,000 jobs. Common sense tells us that if that many employees are gone there will be a real need for training to help those that remain assume their new duties. However, the first round of cuts included 80% of the people in the training function. Evidently, the products and services they were producing were not good enough to ensure their continued employment.

A training function has many processes. Needs assessment is a process as is task analysis, instructional design, instructional development, and instructional delivery. Each of these can be accomplished in several ways. Most training functions have altered the general processes to suit themselves. The rest of this chapter discusses what can be done both to introduce quality considerations into processes and to use them to assure quality in other functions.

DOCUMENTING PROCESSES

The first requirement is documentation. There are three powerful reasons to document processes:

1. Good documentation of process allows training to communicate what will be done and how it will be done.

2. Good documentation allows training to accurately evaluate output, cycle time, costs, and use of resources.

3. Without documentation it's very difficult to modify a process because of the ramifications mentioned earlier.

Describing Processes

Most of the time training describes its processes for itself. With such an internal perspective, training might picture a task analysis like the one in Figure 4.2.

This might work for training, and indeed it might impress customers, but they won't be able to grasp the importance of the process for their own operations. It doesn't speak in language they understand and more importantly, it doesn't meet their needs. However, what if it were described like the one in Figure 4.3?

This chart addresses a customers' potential wants and needs. Customers may or may not understand that the process of doing a task analysis identifies what people do, what

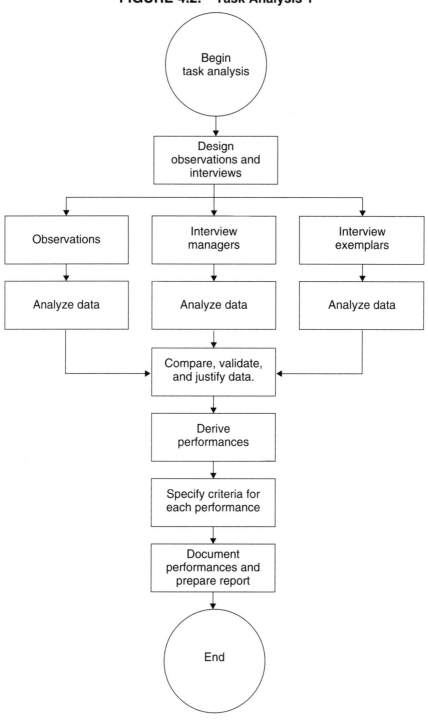

FIGURE 4.2. Task Analysis 1

FIGURE 4.3. Task Analysis 2

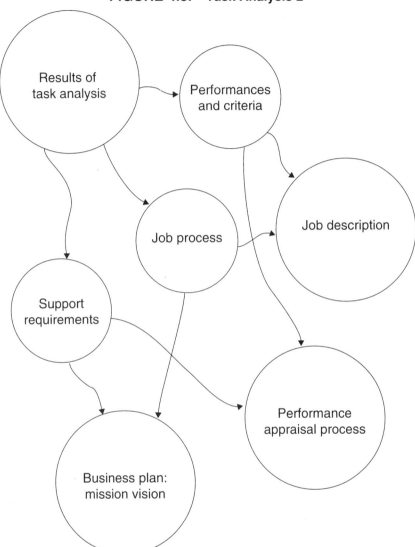

they use to do it (e.g., information, materials, and equipment), and how long it takes. Representing the results of what the tasks analysis discovered in this way shows how that information can be used to help:

- Set performance criteria.

- Write job descriptions.
- Analyze job processes.
- Identify support requirements.
- Ensure jobs are aligned with the organizations mission.

Training's customers must have accurate job descriptions to qualify for ISO 9000 or other quality programs. They must do performance analysis for their employees. The employees themselves will want the training's task analysis to ensure they are being rated against valid criteria. Of course, every training function's charts do not have to resemble the ones provided here. Trainers may choose to simply write it out, but the important point is to direct the message toward customers and state it in ways that address them as customers. Too often trainers look for an audience rather than users.

A major part of the quality movement is being able to describe processes. It is valuable not only for customers but also those within a training function. The act of describing a process allows those engaged in the process to:

- Agree about what the steps are.
- Reach concensus on what steps are necessary.
- Specify what resources (e.g., time, equipment, information) each step of the process and the process as a whole require.
- Evaluate which resources add value, are used well, are redundant, or reduce quality.
- Determine cycle time.
- Agree on what information is required and what is nice to know.
- Come to consensus about how information is used, where it comes from, and what information is missing, late, or incomplete.

An illustration of how documentation of a process can be valuable is shown in Figure 4.4. The illustration shows the test item review process that an IBSTPI task force was developing for an examination to certify instructional designers and developers. The test items required stringent review from experts. Each item has to be approved by separate reviewers at least twice. The task force chairpersons had difficulty visualizing and coming to agreement on the elements of the process. However, after they developed the flowchart, there was no more confusion or need for debate.

A more comprehensive examination of the course delivery and administrative process was conducted by the training function for a division of AMOCO. They began with a process map showing which team members were responsible for different elements of the process. When the process map was complete (see Figure 4.5) each person examined the activities represented by the cells assigned to him or her with the goal of mapping the process and reducing cycle time within each step. When each step had been examined, the team came together to develop a redesign of the process that incorporated all the ideas generated by the process map. In addition to reducing cycle time, the training function also identified redundancy of efforts, omitted activities, more clearly defined roles and responsibilities. They were also able to use the process map when training new people as to their roles. Finally, the map facilitated conversations about skills, abilities, and resources.

Figure 4.6 shows another example from a company that detailed the steps to develop instruction on how to use advertising materials to best advantage. The development process for the instruction is clearly defined. Each participant can trace his or her activities from beginning to end and clearly see what others are doing in their processes. It is easy to understand even for people without much sophistication in instructional development. The second column addresses how the program will be presented. The third is for the instructor. When the entire process is brought together at the end, they will have a video tape, a set of slides, materials for the learners, and a presenter's manual.

FIGURE 4.4. Item Review

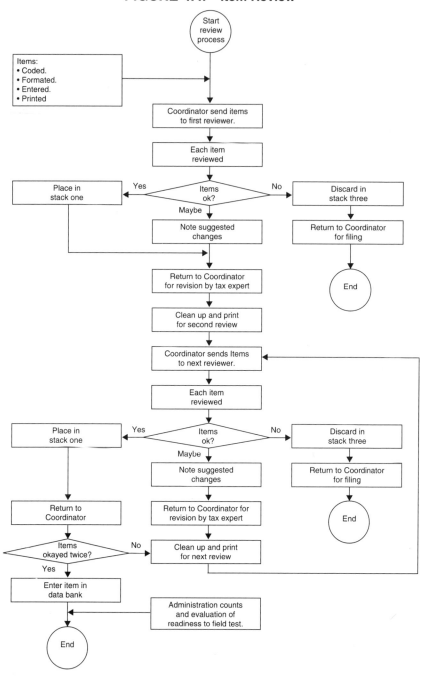

Source: Janet Gregory and Nelle Hanley, IBSTPI Task Force For Certification of Instructional Designers and Developers.

FIGURE 4.5. Process Overview

	Step 1	Step 2	Step 3	Step 4	Step 5	Step 6	Step 7
Team leader		Sign contracts with vendor or contractor					Review evals and enrollments
Registrar		Monitor past quarter's enrollment by class and inform TC of enrollment trends (up, down); schedule room, class dates.	Compile and deliver monthly training schedule, maintain classroom schedule, schedule class in TEAS.	Check enrollment of upcoming course, tell TC; add, delete, or change enrollment status.	Answer student questions about class schedule.	Fill empty spots.	Check, clean up, and close file by month, enroll (remote) students, close remote classes.
Secretary			Order and verify receipt of training materials, ship materials, help ENC; make instructor travel arrangements.	Send reminders about upcoming class, clean up materials, mail prereadings, confirm AV, prepare vendor packets, confirm security, order catering.	Set up hospitality, turn on PCs, put paper in printer, unlock rooms, confirm security has badge for vendor.	Greet vendors, hand over vendor packet and class paper work.	Process and distribute evals to TC and TL, call building and equipment services as directed or needed.
Computer support		Help with the selection of any software or hardware.	Set date and reserve room for PC maintenance, set up PC or LAN for class.	Set up class software for vendor classes (PC and LAN), help instructor.		Be on call.	
Course Coordinator	Initiate or respond, scope job, identify and locate instructor, equipment, classroom, materials and designer, clarify roles, schedule meetings.	Add or delete classes by request, reserve rooms; assign instructor; ask admin to order and ship materials, ask computer support to arrange for and set up equipment.	Coordinate shipments to client; determine if preread materials are needed and ask ADMIN to mail prereadings; give ADMIN information about vendor.	Verify enrollment of vendor courses, coordinate technical and equipment support, cancel vendor courses if necessary.			Review eval summaries, update CAPS, process vendor invoice, prepare and send monthly report to client for total FES costs.
Trainer		Prepare for class.	Confirm materials ordered, ask admin to mail any prereads for own class.	Set up software, check equipment, clean up materials for own course.	Hands our sign in sheet, teach class, if nec. ask ADMIN to contact building services.	Hand out and collect evaluations, summarize evaluations, give roster and evals to ADMIN.	

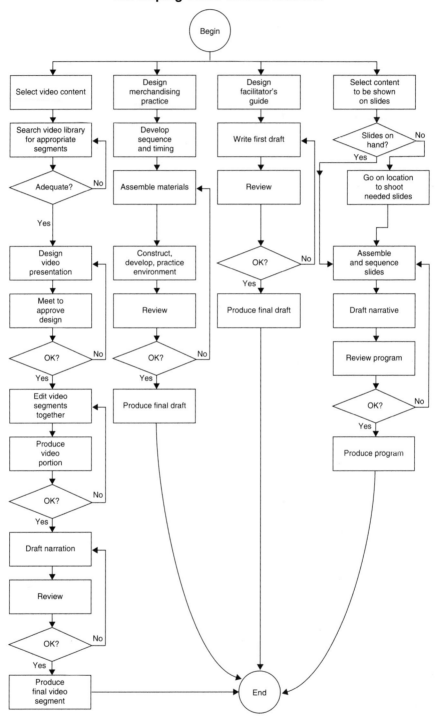

FIGURE 4.6. Process Flowchart for Developing Video Course Material

EVALUATING PROCESSES

The previous section looked at what a process is and how it can be of benefit, this section looks at the requirements for a process. Processes have four requirements:

- Facility and equipment.
- Materials.
- People.
- Time.

To execute a process training must have tools and a place to perform the process; materials and supplies; people skilled enough to accomplish the task; and time. Two of these requirements bear directly on whether quality can be achieved: people and time. If people are unskilled or poorly motivated, quality suffers. Also, if the amount of time is insufficient or too generous, quality also suffers. A process is efficient if it uses skilled people, in the right kinds of ways, and provides personal satisfaction when the work is done well. It is effective if it produces what it is supposed to produce, to the standards set, within a reasonable amount of time.

Efficiency and effectiveness lead directly to quality for the customer. (The caveat is, the customer's wants and needs have been assessed and the purpose of the process is to meet those expectations.) It makes sense then to examine these two concepts more closely.

Efficiency

Efficiency is often associated with economic matters. An operation is considered efficient if it produces products or services and makes a profit. That is a direct reflection of how well people in an operation perform. When workers perform well and attend to their duties faithfully, the chances are good that the organization will make a profit and be efficient. Efficiency also connotes lack of waste. Again, people are the wasters, not machinery or computers. So if the people do not waste, an operation will be efficient.

To be efficient an operation must have skilled people. Not everyone has to be exemplary in their skills, but some should be. Others should have enough skill to avoid waste and loss of time. Unskilled workers can wreck any process quickly and completely. Trainers in particular should be sensitive to this fact because trainers are, after all, responsible for providing the skills. If a process produces poor outcomes, the failure is usually blamed on the workers. However, perhaps the workers should point their fingers at management. If lack of skill leads to poor results, maybe the training function should accept the responsibility.

Management also shares the responsibility. It does not matter how skilled workers are if they're used improperly or not allowed to exercise their skills to the best of their abilities. A manager's job has become primarily the recruitment, development, and assignment of people to jobs for which they are suited. When these things are done well, workers usually produce to the best of their ability without much need for supervision.

Effectiveness

An effective operation produces desired results on time within budget. The emphasis is on the product or service wanted as an outcome of the process. The old definitions of quality control and quality assurance were devoted almost entirely to ensuring an effective operation. For most, therefore, effectiveness is not difficult to imagine. However, current thinking has added perspectives and considerations that were not given much thought before. One such perspective is cycle time.

Cycle Time. Davenport tells us only 5% of the time devoted to a process is spent doing work. The remaining 95% of the time is spent waiting for information, approvals, or feedback.[3] Of course processes require these, but they should not take 95% of the time used. Many processes are, in fact, not designed. They evolve over time. Therefore, they are rarely efficient or effective. Cycle time is made up of the time spent waiting, the time spent moving materials and information, the time actually spent doing something of value, and the time spent examining or checking. Not all of this time adds value. However, it all uses resources and adds cost.

Cycle time is the amount of time it takes to complete one iteration of a process. Cycle time applies to small two- or three-step processes as well as huge complex operations. In each case the mandate is to engineer the procedures used in much the same way a machine is engineered. A good example of how effective this kind of thinking can be is illustrated in the following example.

Fifteen years ago it was accepted that the development of a two and a half hour module of video mediated, self-administered instruction took nine months. The cycle time was nine months. One company would stagger the development cycles to compress time, but they were convinced they could not shorten it except by taking risky shortcuts or working overtime. Today, the same process takes less than four months and the courses the company produces are of better quality than their predecessors. What happened? The company examined cycle time for the process and reengineered it.

Examining cycle time is not difficult. There are four elements:

1. Identify and document the steps of the process.

2. Classify each step as:
   — "A," an action step that contributes directly to achieving the goal of the process such as:
   — Developing interview questions.

---

[3] T. H. Davenport, *Process Innovation: Reengineering Work Through Information Technology* (Cambridge MA: Harvard Business School Press, 1993).

- Conducting interviews.
- Compiling the answers.
- Analyzing the answers.
- "R," a review step where the product or service is being validated such as:
  - Having someone review and approve the questions for the interview.
  - Getting someone to approve the suggested list of people to be interviewed.
  - Checking the accuracy of the analysis.
- "T," a traditional step, which is accepted, but does not add value such as:
  - Waiting to conduct the interviews.
  - Waiting for approvals.

3. Document the amount of time required for each step.

4. Eliminate those steps that do **not** add value to the process or modify them so they consume fewer resources.

Value is added when a step either enhances quality or reduces cycle time. Another way to say it is, whenever customers' needs and wants are met better than before. Customers want the best product or service they can get for a price they are willing to pay at the time they feel the need. Their conception of best product or service defines quality for the provider. Price is relative to quality. People are willing to pay a bit more for a product they think is better in some way. However, what about timing?

If the product or service is not available when a customer wants it, there is no sale. It's that simple. Cycle time is critical, and following is an example for the training function. There are five phases in the full process of developing a training episode:

1. Initiation.
2. Design.
3. Development.

4. Implementation.

5. Evaluation

The initiation phase has four major steps:

1. The customer states the problem or situation they want to deal with. This could be in the form of a request for proposal.

2. The provider (i.e., training function) prepares a proposal.

3. The customer reviews the proposal and either accepts, modifies, or rejects it.

4. The provider and customer agree to a contract.

A flowchart of the process helps visualize the steps (see Figure 4.7).

In this sample flowchart in Figure 4.7, there is an internal department (i.e., customer) asking for assistance from the training function. There are several places where cycle time could be reduced. (Put an A, R, or T in each box or diamond. Ask which steps are required, which ones add value, and which ones are done because we have always done them.) For example, is there a need for a formal contract from an internal department? There should be an agreement stating what will be done, to what specifications, with timelines acceptable to both parties. However, a formal contract may not be necessary. The review steps deserve particular scrutiny. More time is wasted in review than any other activity. Reviews have to be done, but they can be done in hours instead of days. If, for instance, a trainer faxes a copy to the reviewer instead of using the mail, the trainer can save a day each way. One company, for instance, was able to reduce cycle time for the initiation phase from two weeks to two days and deliver higher quality service.

Cycle time delays, of course, appear not only in the initiation phase but in other phases as well. The development

FIGURE 4.7. A Typical Initiation Phase

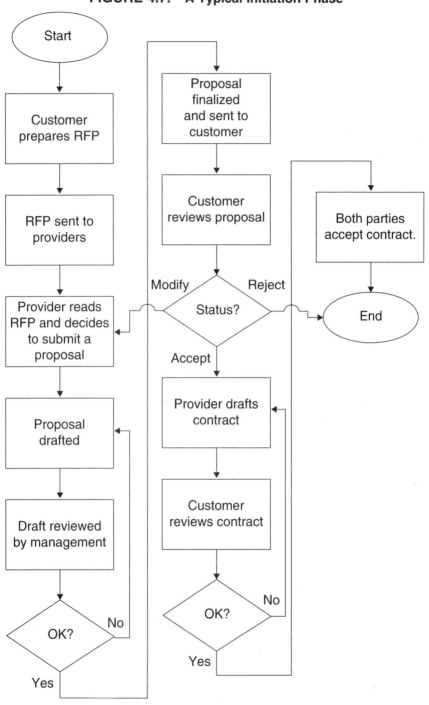

phase, for instance, is rife with wastes of time (e.g, reviews, waits for materials, or procedural snags with subject matter experts). The bottom line is this: Why should a training function take months to develop an instructional episode when they can do a better job in weeks?

In the instructional design process, much time is spent reviewing, redesigning, and redeveloping instruction. Could new standards be developed to reduce this time? In most cases the answer is yes. For example, a course was developed by a famous expert in the instructional design field. The course had six modules. Each module had to be reviewed three times by an expert and then the complete program was reviewed three times. The expert made, mostly cosmetic, changes to each draft, returned it, and insisted upon seeing the corrections before he authorized continuing with the project. He conducted more than 20 reviews. In addition, the materials were reviewed by internal QC, content experts, and management. In all there were more than 30 reviews. The course contained 15 hours of instruction. Of the 18 months it took to develop the course, twelve of those months were used for reviews. Although this may seem to be an extreme example, it is **not** uncommon. Trainers can usually make drastic savings in time simply by examining each review, how it's done, and how long it takes.

A typical review of instructional materials proceeds in five steps:

1. A draft is printed and packaged (half day).
2. It is sent to the reviewer (one day in transit).
3. The reviewer performs the review (two days).
4. It is returned to the developer (one day in transit).
5. The developer makes suggested changes (two days).

Steps 3 and 5 add value. The others are two and a half days of wasted time. Usually the time needed for this review (even in steps 3 and 5) can be reduced significantly. A reviewer simply works with the developer at the developer's work station reviewing, suggesting changes, agreeing on what should be done, and doing it. A process that once took a minimum of seven and a half days can easily be reduced to two. And there is a bonus. Since the two people are working together, they have more insight about the finished product.

Relationship Maps. There is another easily achieved way to reduce cycle time. A relationship map is a tool that can be developed for detailed study of any process. It is used to curtail in-process redesign and avoid rework. In-process redesign occurs when someone encounters a problem during a process for which there is no clear-cut solution or procedure. They are forced to create a solution as the process continues. This sometimes leads to grievous errors and usually slows progress, wasting time. The problem of rework crops up when a product or service produced as an outcome does not meet standards set for it by internal control or the customers. Rework costs both dollars and customer confidence.

A relationship map is shown in Figure 4.8. This one was developed to show the relationships for a training function production process. This is the first view. The second views would be detailed maps for each subprocesses (i.e., master and print, assemble, bind, and quality checks). The advantages of this procedure are that it:

- Is a visual representation of the process.
- Allows flexibility in the performance of tasks.
- Provides easily understood constraints about what should happen when and where.
- Only takes a single sheet of paper (four for the second views).

Resource Use. Cycle time is an important item most training functions should address immediately. Use of resources, however, is something most training functions have been

FIGURE 4.8. A Relationship Map

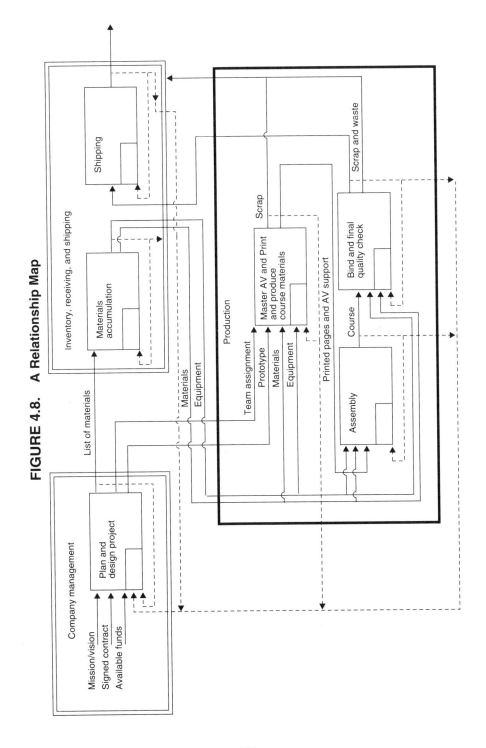

working to control for years. Resources include facility, equipment, materials, and people. They are the elements of a process that can be obtained in some way. Most managers spot wasted materials much easier than they do wasted time. Of course, at times, conservation of resources can have negative effects. For instance, working in an inadequate facility or a department lacking support staff can be counterproductive. Therefore, an important measure of a process' value is how it uses resources. Once a process is described, managers and employees can ask questions about what resources are being used. How, when, and by whom information is accessed, used, stored, and sent to others. How cycle time and costs would be affected if someone else accessed, used, and sent information. How cycle time would change if a different technology were used. The training function should continue to pay attention to resources.

CONCEPTUALIZING AND USING PROCESSES

Obviously, there is value in processes and importance in examining them to make sure they function efficiently. This section examines four major processes performed specifically by the training function. Often, for one reason or another, these processes hardly ever get formalized and documented. A detailed discussion can give some insights about how to develop a process and use it to an advantage. These four processes are:

- Evaluation.
- Course maintenance.
- Administration.
- Developing training function personnel.

The Process of Evaluation

One is the process of evaluation. It, like many others, is assumed. Because of this assumption, it's often not done well or is skipped altogether. But, particularly when quality is the goal, it's critical to smooth operation of a training function. Figure 4.9 presents the total process for developing an instructional episode.

Even though the final step in this process is evaluation, evaluation actually happens in each step and between formative steps. Of course, the summative step called evaluation must include both formative and summative measures. Before an examination of where evaluation occurs throughout the instructional development model can be explored, a few terms must be defined. Two of the terms are *formative evaluation* and *summative evaluation*. They are essentially the same as formative and summative feedback. Formative evaluation is internal to a process. Summative evaluation occurs when a phase has been completed. An *audit* is an activity that takes place downstream to decide if and how much transfer of learning has taken place, are the skills learned in implementation actually being used on the job, and if not, why not? A *final analysis* is often taken for granted, but it, too, should be planned for. It provides a historical perspective; someone, sometime must decide whether the effort was successful in terms of the organization's mission or vision.

Figure 4.10 provides an idea of where in the overall instructional development process each type of evaluation should happen and the form it may take in each phase of a process.

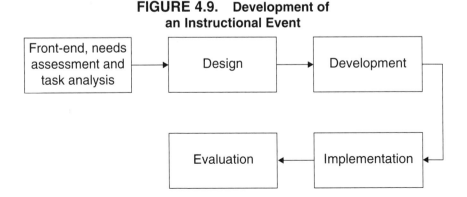

FIGURE 4.9. Development of an Instructional Event

FIGURE 4.10. Evaluation Opportunities

Type of Evaluation	Front-end Activities	Design	Development	Implementation	Evaluation
Formative	Document reviews, QA plans, data analysis.	Developmental tests, subject matter expert reviews, internal reviews.	Internal reviews, developmental tests, subject matter expert reviews.	Observations, internal reviews.	Comparison of design specs with actual internal reviews.
Summative	Customer comparison with wants.	Acceptance review by customers.	Pilot tests, field tests.	Customer feedback, exam results.	Analysis of pre-post test results.
Audit	Design audit trail.			Comparison of actual delivery with design.	Downstream evaluation on the job.
Final analysis	Parameters set.			Data preserved.	Judgement made and documented.

The process of evaluation will be looked at more fully in Chapter 6. What's important to remember here is evaluation is a process which should be not overlooked by the training function.

Course Maintenance

Course (or curriculum) maintenance is another often overlooked process of a training function. If an instructional episode is to retain value, it must continually serve an appropriate audience through appropriate content (and skills) and delivery mechanism. It should maintain its instructional integrity throughout its tenure as an active course. Figure 4.11 provides a model of what a process for maintaining an instructional course might look like.

Every course maintenance process need not look much like the one in Figure 4.11. Whichever way a course maintenance process is defined, it must be documented, and the course evaluations must occur on a timely basis. As the needs of an organization change, so will the courses the training function offers. If training is to add value, it must reflect customer's current needs and expectations. The benefits of documenting a process for maintaining courses are that it helps:

- Identify what training resources to commit.
- Compare the cost of maintenance against the cost of replacement.
- Negotiate and schedule resources from customers.
- Identify opportunities to shift old courses to a different delivery format such as computer-based training or CD-ROM.

Administration

Another process performed by a training function is that of administering a single course or instructional event. Although

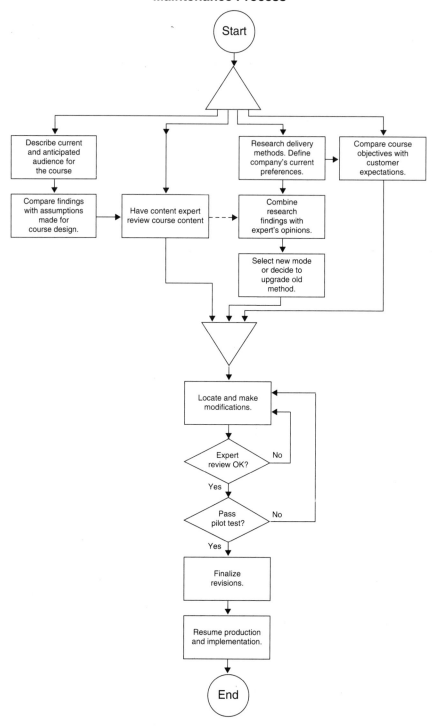

FIGURE 4.11. Course Maintenance Process

this may seem secondary to the actual act of design and delivering a course, much wasted time and resources can be traced to poor or missing administration.

For instance, much preparation is needed to conduct a classroom experience. The room has to be found and prepared. Materials have to be duplicated and assembled. The instructor has to be selected and trained. During the course, students must eat, sleep, and attend class. When problems occur, someone should be available to work them out. After the experience is completed, student scores should be documented somehow and records of their attendance and accomplishments kept. Also, documentation should be made available to the learners and their management. Procedures should be in place to allow learners to question decisions made contrary to their best interests. Finally, action may be required to comply with the Americans with Disabilities Act and avoid racial, sexual, or ethnic bias.

All these activities are administrative, though many of them are performed by instructional professionals. Well-documented procedures for them and, perhaps, the whole of administration for the function, provides two things: an informative job aid for people who do not do these things very often and a standardized process that makes outcomes more predictable.

There are many administrative problem areas to consider. One example involves the administration of self-directed learning for people located far from the central location. Figure 4.12 shows one possible process. Not all the activities for this process are administrative, but many of them are.

Many processes are filled with administrative tasks. Sometimes they are justified, sometimes they are not. Documenting the process makes it much easier to make time and resource-saving modifications. For example, one company had three full-time staff members assigned to handle course registrations. Management asked why the process required three people. Training documented the number of transactions required to register students. It discovered that for every three registrations another two transactions were required to han-

FIGURE 4.12. Distributed Self-Study Education Administration

Administrator	Coordinator	Learner
Market, send catalogues, etc.		
		Select learning experience.
		Prepare and mail application.
Process application.		
Print and package materials.		
Assign coordinator.		
	Prepare for assignment.	
Mail materials.		
		Examine materials.
		Complete first session.
		Mail work to coordinator.
	Read and critique.	
	Mail critique to learner.	
		Accept and read critique.
		Work on next session.
	Critique final session.	
	Evaluate learner.	
	Mail to learner and inform administration.	
		Accept results.
		Return materials.
Document results.		
Inform learner and learner's manager.		

dle cancellations, no shows, and moving people off the wait list.

By documenting the actual number of transactions, training is in a better position to compare the cost benefits of:

- Outsourcing registrations.
- Investing in a different electronic system that allows students to self-register.
- Reclassifying the job and hiring personnel at a lower salary grade.

Developing Training Function People

One final process deals with internal staff—specifically, the process used to develop training function staff. ISO qualification requires staff members to receive development. Figure 4.13 provides a possibility. This figure shows only a subprocess. Such activities as budgeting, administration, and evaluation, while not shown, are a part of the larger process.

The benefits to documenting the maintenance and development of training personnel's skills are it allows you to:

- Identify and schedule individual's developmental activities to match what is required to service customers.
- Communicate the process to staff and customers.
- Explore other more cost-effective ways to develop staff.
- Document the cost so training can later:
  — Measure the return on that investment.
  — Compare the cost benefit of development against outsourcing.

SUMMARY

When training fully understands the processes it uses and how it contributes to cost and cycle time, it can begin to chal-

FIGURE 4.13. Training Staff Development Process Checklist

Outputs	Process activities	Time	Cost	Done
Performance reviews conducted.	Managers and supervisors meet with staff.			
Individuals agree with results.	Survey staff through personal interviews.			
Personal wants of staff confirmed.	Survey staff through personal interviews.			
Organization's needs projected.	Analyze latest needs assessments.			
	Survey customers (phone and in person).			
Skill inventory.	Analyze performance review data.			
Select development opportunities for function.	Research what is available internally and externally.			
Projected skill needs of function over next 18 months.	Compare skill sets in hand with projected needs of organization.			
Match staff with skill development opportunities.	Compare job duties of each person with the organization's needs.			
Development plan.	Meet with individuals to schedule development.			
	Document agreements.			
Progress report.	Survey staff each quarter to check on progress.			

lenge (ask hard questions) about what services it should be providing (because the function is cost competitive), what services it should outsource, and what services would be better provided through strategic alliances with others. Training can ask questions about how time and cost would be affected if it changed vendors, trained staff, made use of better technology, or simply got rid of the unnecessary, non-value-adding activities. In short, if training has a thorough understanding of its processes, it can steadily provide better quality at lower costs.

Check Your Understanding

Use the checklist that follows to check how well you understand how to evaluate **your** processes. Check the appropriate column if you can answer yes or no. Use the comments column if you aren't sure and to record any remarks you might have about cost, cycle time, and standards for your input, output, and outcomes.

Question	Yes	No	Comment
1. Are your key processes documented and described so they can be analyzed?			
2. Can you identify the points in your processes that add unnecessary cycle time?			
3. Can you identify steps in your processes that add value and those that do not?			

Question	Yes	No	Comment
4. Have you used your core training processes to facilitate discussions about costs, time, and resource use? About how staff are assigned? About how you cost jobs? About where your costs are not competitive? About what processes are currently effective? About what processes are currently efficient?			
5. Are you in a position to evaluate the effectiveness and efficiency of your core processes? What services do you currently provide competently? What services must be brought up to a higher standard?			
6. Have you documented your course management process?			
7. Have you documented your administrative processes?			

What to Do Next

Review your answers to the questions on the checklist. Ask each person who is directly concerned with training (including key customers) to do the same. Then meet with them to make these decisions:

1. Are there processes that should be redesigned or reengineered?

2. Are there processes that seem to be missing and should be developed?

3. How can we decide beforehand the nature of each process as it will be after revision or development?
4. Which measures are most critical, what are our priorities?
5. Do we have the expertise, time, and resources in-house to do what should be done? How much of it? To what standards?
6. Where can we find the expertise we require?
7. How will these activities fit into the mission, vision, and day-to-day activities of our parent organization?
8. What are the critical steps required to begin?

5
TRAINING AND STANDARDS

✸ *The peregrine falcon is a special bird. But could you recognize one if you saw it? Probably not. You know it's a bird of prey. You know it attacks other birds in midair. You know it nests on cliff sides or skyscrapers. So, if you saw a raptor catch a duck in midair, you might think it was a peregrine, but how can you be sure?*

It would help to know how big they are, how fast they fly, what colors they have and how those colors are distributed. It would help to know where in the world to expect to see one (besides Chicago). It would help to know how they move and attack in the air. It would help to know how they mate and feed their young. It would help to be able to compare a peregrine with other birds of prey like red-tailed hawks, ospreys, or eagles. In short, you need a better description of the bird.

Let's pretend for a moment that you are the city planner for London, a city plagued by pigeons. You decide to introduce peregrines to that environment by bringing in six pairs of them. You contact a naturalist, who can provide the birds at your request. She invites you to come to her wilderness setting area to select the pairs you want. When you arrive, she takes you to an isolated area where, with the aid of binoculars, you can see falcons flying, nesting, and preening. Which pairs will you choose?

If you see one dive toward a duck in flight and pull up short or miss it altogether, you may decide not to choose that peregrine. Is that a wise decision? How often do they miss? What does a "good one" look like? If you see another make a clean kill, will you choose it? Given the choice between the one that missed and the one that succeeded, you probably would. The peregrines, themselves, have helped you make the decision. They have provided evidence of competence you can use. They have given you a standard for comparison.

The same thing is done when establishing standards for a training function. This is particularly true if the training function is involved with its organization's movement toward total quality.

The criteria used to judge quality get more stringent everyday. As has happened with other products, competition is helping customers refine their training requirements. For example, a trainer does something special, then explains to a customer what was done and how it worked. This customer now has an idea of how other training might be improved. The next vendor the customer uses will probably find a requirement that the new training contain the attribute that made the former effort a good one. The new requirement became part of the standard the customer uses to judge the acceptability of training.

This chapter explores the issue of standards as they relate to the training function. It examines:

- The purpose of standards.
- Uses of standards.
- The standards currently available to training.
- How you might evaluate your standards.

THE PURPOSE OF STANDARDS

Standards help customers compare products and services. They are what leaders use to communicate what they expect. Standards allow everyone to judge the performance of people and what those people produce. Standards are what self-directed teams use to guide their work.

But what are standards and, more specifically, what about standards for training? A starting point is a look at a basic definition of the word. Webster's Collegiate Dictionary defines a standard as, "Something established by authority, custom, or general consent as a model or example: criterion." The key words in the definition are model, example, and criterion.

Models and Examples. In everyday life, people are always applying models and examples. A model could be as simple as a handcrafted rendition of a product someone intends to manufacture. A model could be someone's idea of what would be a better-than-average item selected from a group. An example is a boy examining a handful of marbles. He says, "This is the best aggie of the bunch." He has selected a model, or standard, he will use to judge the adequacy of the rest of the marbles. When people see a gymnast achieve a perfect score, they can use that performance as a standard. From that point on, they will be able to compare other gymnasts' performances to the perfect one. The critical element lies in a person's ability to identify what makes the model or example good. The boy who made the statement about the marble must be able to convince his friend the chosen one is best. He must point out the attributes that make the marble a paragon. Similarly, there are elements of a gymnast's performance that cause judges to accept or reject it as exemplary. The gymnast must land on both feet, must maintain balance, and complete the routine as it was described to the judges. Models and examples become standards when they meet certain preconceived notions about how they should be followed and used. They are standards because customers accept them.

A customer's previous experience becomes the standard used to judge the adequacy of future training. For example, if previous training had tests, customers will expect tests they can use to discover if students learned what they were supposed to learn.

Criteria. Whether the standard is a model or an example, it communicates a set of criteria to be used to make a judgement. In effect, those who decide on the quality of a product or service do so by comparing it to something. The comparison is done by use of criteria. The criteria may be explicit or implied.

How Standards Come to Be. Webster's definition of a standard also referred to methods used to define a standard:

"... established by authority, custom, or general consent."[1] Most standards discussed in this chapter have been developed by an acknowledged authority, be it a professor, a skilled practitioner, a regulatory agency such as the U.S. Department of Labor, or a qualifying agency such as ISO or IBSTPI.

There are other, less formally stated ways that standards come to be. They can evolve through custom or general consent. This chapter will first look at customs and general consent before looking at formal standards for training. When a standard is developed through custom, there can be differences about whether the standard is applicable. For example, it's customary in many organizations for training to be presented in a classroom by an experienced presenter or trainer. General consent provides that learners should appreciate the experience and that it shouldn't impose on their moral or ethical values. Although these kinds of considerations may or may not contribute to the productivity and health of an organization, they are real and must be acknowledged. The standards used by learners to judge quality are not necessarily the same as those used by management for the same purpose.

What an average learner expects may or may not be valuable for learning, but it is what is expected and, therefore, is a standard that will be used to judge quality. If a training session, for instance, doesn't take place in a classroom it may be rejected in the minds of some learners because it doesn't meet their expectations. They won't be able to apply the standards they are familiar with to the session they are experiencing. Those standards cover such factors as how the instructor should look, talk, and act; how the classroom should be set up; the condition and physical appearance of handouts and materials; and the equipment used. It doesn't matter whether these elements are germane to the purpose of the training. Because the learners expect to see them, a certain amount of dissonance is caused by their absence.

It's very important to remember that learners are not usually the primary customers. They may be the people for

---

[1] *Webster's Ninth New Collegiate Dictionary* (Springfield MA: Merriam-Webster).

whom the training was designed and developed, but they aren't the people who determine if it is needed, nor are they paying for it. In effect, there are two sets of customers. One set, the learners, are customers for *learning*. The other set, management, are customers for *training*. Learning is how individuals acquire knowledge and skills. Training is acquired by an organization to augment and upgrade the performance of production units or service efforts. Because learning and training are not the same, the standards used by learners to judge a training episode usually are not the same ones managers use to judge the quality of the same training.

It's important for a training function to know about these differences and take them into account. For example, if a new program is to be offered through computer-based training, a training function must prepare the way and give learners reasons for using the new medium that include standards used to judge whether it makes for a valuable experience. Many programs have failed simply because this sort of preparation was ignored or done poorly.

Uses of Standards

The training function can use standards for a variety of purposes. Federal Express (FedEx) believes in standards. For example, it certifies its trainers based on standards. FedEx also uses standards to prepare supervisors. The training department's award-winning program, LEAP, screens and develops potential supervisors based on those standards. FedEx also uses computer-based training (CBT) to deliver its training to employees worldwide. Standards were used in the development of its CBT.

Other companies have developed standards for material. Amway, for example, has standards for course content: how to sequence it and how to format it. They developed these standards to reduce instructor preparation time and translation costs.

Testing agencies like Educational Testing Services (ETS) use standards to certify instructors. IBSTPI uses standards to certify training functions, the process used to qualify instructors, instructional designers, and instructional products.

FedEx started the ISO certification process in 1993 and received ISO registration in 1994. What is unique about their effort is they asked that all operations at all locations be registered. The fact that FedEx has documented standards and uses those standards contributed significantly to its success. Also significant is that a renowned service organization saw the value in ISO certification—particularly instructive for training.

In fact, ABB's major training centers in Ohio, New York, and Sweden are all ISO registered. The uniqueness of these registrations is they are training, not manufacturing, facilities. To achieve registration, ABB documented standards for all the training functions at those facilities (i.e., design, development, production, administration, delivery, and evaluation). Those standards provide a firm foundation for ISO evaluation and continuing growth of the training function and its influence within the company.

Standards are used by training managers to help them manage their departments. Standards communicate what is important. They help make decisions about what to do. They let other people know what is expected. Perhaps the greatest value of standards is in the process of coming to agreement about what they are, how they will be used, and who should define them.

Training is basically a service, not a product. Yes, there are products associated with it such as student guides, instructor manuals, and audio-visual aids. The critical aspects, though, are the activities and information that promote learning. To identify what activities and information are needed to support learning, trainers do essentially four activities:

1. *Preparation and qualification; what trainers call front-end analysis.* Trainers prepare to design and develop materials by assessing the needs of customers and the target audience. They analyze tasks to discover the knowledge and skills required by practitioners. They find out

who the learners will be and how they might learn efficiently. They develop criteria used to judge success when training is complete.

2. *Design and development.* Trainers make decisions about what kind of experience to provide the learners and how it should be carried out. They develop the materials needed and test them before involving the learners.

3. *Implementation.* Trainers provide the service. They conduct the training, administer it, and respond to problems and situations.

4. *Evaluation.* Trainers and others develop ways to discover how well the training was received and how well learners achieved the criteria set during preparation. They report these findings to their customers.

Each of these are part of a larger process. Each phase of the training process is examined in terms of how standards can be used.

Inputs. Inputs refers to the information training relies on to make decisions about what customers require at the organizational, job, and individual levels. At the organizational level, inputs are the customer's business drivers, anticipated regulatory changes, and market factors. At the job level, inputs are the skills and knowledge required to perform a task (i.e., the content of a course). At the individual level, inputs are a person's performance deficiencies and developmental needs. Training gets its inputs when it performs needs assessments, task analyses, and individual skill assessments (i.e., the front end analysis activities). Training can use standards to communicate what it has to have to anticipate customer requirements and deliver cost effective interventions. Standards can address:

- Who to involve to ensure people with the appropriate information and perspectives are included.

- How to design studies to ensure validity of the results.
- How to conduct assessments.
- How to report findings and recommendations.

Figure 5.1 is a sample of a standard that relates to inputs.

Processes. The processes training uses are the assessment, instructional design, development, production, delivery, administration, and evaluation processes. Standards at this level define what is to be done, why, and how it will be judged in the end. Training can use these standards to communicate what it requires to control costs and reduce cycle time. Standards include:

- How to select personnel (i.e., instructors, instructional designers, product vendors, and consultants).
- How to evaluate people (instructors, vendors, and designers) and product (specific interventions) performance.
- How to select media and delivery formats.
- How to design instruction:

FIGURE 5.1. Sample Input Standards

Needs assessments will be conducted for training projects. The assessment process will be documented and will include:

- The objectives.
- Representatives from all stakeholders.
- Adequate samples to ensure representation and validity of the results.
- A sufficient number of data gathering methods to ensure validity of results.

— How to select instructional strategies and tactics.
— How to sequence instructional events.
— What instructional elements to include.

- How to evaluate a process in terms of cycle time and cost.
- How to design print and electronically-delivered materials.
- How to implement a course or service.

Figure 5.2 gives an example of process standards. This set of standards is typical of the sort of criteria used with most "soft" skills or activities. It provides guidance rather than specifications, very much like the standards used by editors to guide writers.

Products. These are the print, visual, and electronic materials training produces. It includes the curriculum and course catalogues. Product standards are ones that deal with print formats and guidelines for what to include in administrative guides.

Outputs. These are what training delivers. They include:

- How many times a course is offered.

FIGURE 5.2. Process Standards

The process used to develop instructional materials will be documented. The documentation will include:

- The activities required.
- When and by whom activities are performed.
- What and when information sources are required.
- A rationale for its design and resource requirements.

- How many days an instructor delivers training.
- How many hours of training people complete.
- How many hours a facility (rooms and carrels) are to be in use.

Saturn's requirement that everyone receive a minimum of 92 hours of training a year is a standard.

Outcomes. This is the effect of training's efforts on the organization. Training can use these standards to measure and communicate:

- What new skills and knowledge learners acquired.
- What new skills and knowledge learners applied on the job.
- What content learners retained.
- What level of competence learners achieved.
- Whether or not learning was transferred to the job.
- Whether or not the business requirement was satisfied.

Figure 5.3 is an example of a standard related to outcomes. Standards allow trainers to communicate more clearly what is important to both the training function and its customers. When trainers begin to evaluate the effectiveness of a course

FIGURE 5.3. Outcome Standards

> Cost benefit analysis will be conducted for programs requiring an investment of $50,000 or more. There will be a documented process for getting feedback from customers to determine the value of a program. The measures of the value will be based on customer, business, or regulatory requirements.

Training and Standards 129

or even the training function, standards give the target to which performance can be compared.

Available Standards

Quality standards for each of these phases can be found in several credible sources. Federal agencies and departments, for example, have set standards for training developed for export to emerging nations and others. This chapter looks at three sources that provide standards for a training function:

- ISO Standards.
- U.K. Training Standards.
- IBSTPI Standards.

ISO Standards. ISO 9001, 9002, and 9003 address similar training standards for an entire organization. The standards are about when, where, and for whom training should be provided. They refer to ensuring vendors are qualified, people who perform quality control and assurance activities are qualified, executives are qualified in their understanding of standards, and training needs are identified. All of this must be verified and documented for the inspection group. A simplified version of ISO requirements is given in Figure 5.4

The training function's role is to develop an understanding throughout the organization about ISO, its requirements, and its benefits to the company. Training usually develops the processes to confirm whether employees and suppliers are qualified. Training determines who must be trained or retrained. (As you look at the ISO standards in Figure 5.5, think about how you might meet these standards in your own training function.)

U.K. Training Standards. The British government confronted the need for training standards in the 1980s. In 1990 it began to implement its own program called "Investors in

FIGURE 5.4. Summary of ISO Training Requirements

What Types of Training Programs are Required?

The facility shall implement a training program to ensure that all personnel can carry out its duties in a way that is consistent with the objectives of the quality sytem. This documented program shall:

- Identify skill shortages by means of examination or other techniques.
- Secure the appropriate training resources.
- Implement the training.
- Verify training effectiveness by means of examination or other techniques.
- Conduct post-training monitoring, as appropriate. Appropriate records of training and competence levels shall be maintained for each employee.

Source: *ISO 9000: The Standard for World Class Quality, An Executive Overview* (Southfield, MI: Perry Johnson, Inc.).

People."[2] The U.K. standards are grouped under four headings:

- Commitment.
- Planning.
- Action.
- Evaluation.

They are presented as principles and assessment indicators. The section for commitment appears in Figure 5.6.

---

[2] G. Wallbridge and D. Weaver, " 'Investors in People': A UK Standard for Training and Development," *ASTD International Conference and Exposition* (ASTD: Fairfax VA, 1994).

Training and Standards

FIGURE 5.5. ISO Training Standards

Book and Section	Standard
	Personnel:
ISO 9001 and ISO 9002 Section 4.1.2.2	The supplier shall . . . assign trained personnel for verification activities . . . carried out by personnel independent of those having direct responsibility.
ISO 9003 Section 4.1.2.2	The supplier shall . . . assign trained and/or experienced personnel for verifying that product conforms to specified requirements.
	Training:
ISO 9001 Section 4.18 and ISO 9002 Section 4.17	The supplier shall establish and maintain procedures for identifying the training needs and provide for the training of all personnel performing activities affecting quality. Personnel performing specific assigned tasks shall be qualified on the basis of appropriate education, training, and/or experience as required. Appropriate records of training shall be maintained.
ISO 9003 Section 4.11	Personnel performing final inspection and attests shall have appropriate experience and/or training.
ISO 9004 Section 18.1	The need for training of personnel should be identified and a method for providing that training should be established. Consideration should be given to providing training to all levels of personnel within the organization. Particular attention should be given to the selection and training of recruited personnel and personnel transferred to new assignments.

(continued)

The commitment section of the standards typifies the others in format and function, yet it is somewhat unique. Although other sets of standards imply what is expected of management, few of them are specific and comprehensive in their treatment of employees.

FIGURE 5.5. Continued

Book and Section	Standard
Executives	Training should be considered that will provide executive management with an understanding of the quality system together with the tools and techniques needed for full executive management participation in the operation of the system. Executive management should also understand the criteria available to evaluate the effectiveness of the system.
Section 18.1.2 Technical & Production Personnel	Training should be given . . . to enhance . . . contribution to the success of the quality system. Training should not be restricted . . . but include assignments such as marketing, procurement, and process and product engineering. Particular attention should be given to training in statistical techniques, such as process capability studies, statistical sampling, data collection and analysis, problem indentification, problem analysis, and corrective action.
Section 18.1.3 Supervisors	All . . . supervisors and workers should be thoroughly trained in the methods and skills required to perform their tasks.

Source: ISO 9000: Handbook of Quality Standards and Compliance (Englewood Cliffs, NJ: Prentice Hall).

IBSTPI's **Standards.** Perhaps the most comprehensive and useful standards are those developed by IBSTPI. They describe what an organization, a training function, and training professionals are expected to do and accomplish. IBSTPI's standards are similar to those developed by the U.K. They specify performances and criteria instead of principles and assessment indicators. And most importantly, the performance levels they specify can be measured using the supplied or inferred criteria. IBSTPI has developed standards for:

FIGURE 5.6. U.K. Commitment Training Standard

"An 'Investor in People' makes a public commitment from the top to develop all employees to achieve its business objectives."

Principles

- Every employer should have a written but flexible plan that sets out business goals and targets, considers how employees will contribute to achieving the plan and specifies how development needs in particular will be assessed and met.
- Management should develop and communicate to all employees a vision of where the organization is going and the contribution employees will make to its success involving employee representatives as appropriate.

Assessment Indicators

1.1 There is a public commitment from the most senior level within the organization to develop people.
1.2 Employees at all levels are aware of the broad aims or vision of the organization.
1.3 There is a written but flexible plan which sets out business goals and targets.
1.4 The plan identifies broad development needs and specifies how they will be assessed and met.
1.5 The employer has considered what employees at all levels will contribute to the success of the organization and has communicated this effectively to them.
1.6 Where representative structures exist, management communicates with employee representatives a vision of where the organization is going and the contributions employees (and their representatives) will make to its success.

- The management of the training function.
- The process for qualifying instructors.
- Training managers.
- Instructional designers.

- Training products and programs.

Within these are standards for inputs, processes, products, outputs, and outcomes. Selected standards are reproduced in the next paragraphs because they provide a comprehensive model for other efforts. Obviously, all of the standards can't be reproduced. However, it is worthwhile to examine the entire set of standards and adapt them for use in your organization. One important issue to keep in mind is that standards aren't very useful if they are difficult to measure. The criteria in the IBSTPI examples of standards can be used to discover the presence and adequacy of a performance. (As you read the standards visualize or conceptualize how they could be measured in your organization.)

Standards for the training function. IBSTPI has seven categories of standards for the training function (see Figure 5.7). They are based on the following five principles:

Training:

1. Operates as a well-run business whose business is learning and performance improvement.

2. Adds value through an appropriate range of services depending on the needs of the host organization.

3. Provides quality products and services, on time, and within budget.

FIGURE 5.7. IBSTPI's **Categories of Standards**

• Customer Relationships.	• Leadership.
• Standards and Measures.	• Processes.
• Performance.	• Change.
• Resource Use.	

4. Supports the achievement of its parent organization's mission, objectives, key initiatives, and business strategies.

5. Documents its processes so they can be shared, managed, and improved.

Each standard is supported by specific performance levels and criteria. Chapter 2 presented the entire standard for leadership: rationale, performance, and criteria. The remaining standards shown in Figure 5.8 contain only the rationale. The performance levels and criteria for the standard on standards and measures are shown in Figure 5.9. (Consider how well your training function could meet these standards.)

Standards for the training manager, instructional designer, and instructor. Like the standards for the training function, IBSTPI's other standards are very detailed, containing performance statements and criteria. Figure 5.10 lists the number of performances and criteria for each.

Figures 5.11 through 5.13 list the training manager, instructional designer, and instructor performance statements. (Identify those that would help you communicate what your training function does and how it does those things.)

These samples of standards serve to emphasize the growing importance of the training function. In light of the quality and reengineering movement, where training takes an increasingly more prominent role, applying standards to the training function holds training just as accountable for its performance as any other function.

SUMMARY

Not long ago, organizations wouldn't have been as concerned with the standards of the training function; they would probably not have specified roles and responsibilities for training or trainers. Today, those roles and responsibilities are ex-

FIGURE 5.8. IBSTPI
Rationale Standard

Standard	Rationale
Customer Relationships	**Purpose:** To ensure that you (the function), how you operate, and what you produce reflect the perspective, values, and needs of your organizational customers. **What:** Your relationship with your customers is that of a business partner. You incorporate the voice of your customers in your goals, plans, and measures. **Why:** So you have access to the information (data, interpretations, decisions, commitments, actions, etc.) needed to more fully understand and anticipate the learning and performance improvement needs of your customers. So you can make better decisions about resources you need and how to best use them.
Standards and Measures	**Purpose:** To ensure what you (the function) do, how you do it, what you produce, and the results you achieve are timely, cost effective, and contribute to the accomplishment of your customers' goals. **What:** You evaluate the performance of your processes, products, people, and suppliers. The factual basis for your evaluation includes customer satisfaction, differentiation from competitors, deviation from standard, financial performance, and business results. **Why:** To ensure your decisions are based on facts and reliable data.
Processes	**Purpose:** To ensure that you (the function) are efficient and effective, and your products and services add value to customers. **What:** You design, measure, and manage the processes used to identify needs, select and develop products and services, and evaluate results. **Why:** So you are proactive and responsive, and efficient and effective in your operations and implementation of services.
Performance	**Purpose:** To ensure that you (the function) are a supplier of choice by your customers. **What:** You operate the function as a cost competitive, responsive, timely, and efficient business that provides products and services valued by customers. **Why:** To ensure customers select you as a provider of learning and performance improvement products and services.
Change	**Purpose:** To ensure that you (the function), through your leadership, and products and services facilitate performance improvement in your customers' organizations. **What:** You support the creation of and are an advocate for the implementation of a change strategy in collaboration with your customers. **Why:** So you can add value and support your customers' business goals.
Resource Use	**Purpose:** To ensure that you (the function) have the resources (people, facility, equipment, systems) you need to provide products and services your customers value. **What:** You identify the resources required to support your function (for example, people, facilities, equipment, systems, etc.). You decide how to best secure those resources. You manage how to best use those resources. **Why:** So you have available the most appropriate and cost effective resources when needed.

Source: reprinted with permission of the International Board of Standards for Training, Performance, and Instruction (Barrington, Illinois). © IBSTPI, 1994.

FIGURE 5.9. IBSTPI Standards for Standards and Measures

Rationale	Performances	Criteria

Purpose:
 	
To ensure what you (the function) do, how you do it, what you produce, and the results you achieve are timely, cost effective, and contribute to the accomplishment of your customers' goals.
 	
What:
 	
You evaluate the performance of your processes, products, people, and suppliers. The factual basis for your evaluation includes customer satisfaction, differentiation from competitors, deviation from standard, financial performance, and business results.
 	
Why:
 	
To assure your decisions are based on facts and reliable data.
	1. Standards and processes exist to evaluate staff, supplier, product, process, service, and overall function performance.
	2. A process exists to benchmark key processes.
	3. Staff, suppliers, and customers know the goals, standards, and measures used to evaluate individual, product, process, and overall function performance.
	4. The standards, measures, and metrics are documented and validated.
		1. The voice of your customers and suppliers is incorporated in setting your standards and measures.
		2. The metrics include:
		  – Financial measures (ROI, ROE, etc.).
		  – Customer satisfaction.
		  – Deviation from standards.
		  – Cycle time.
		  – Cost per development hour.
		3. Data derived from benchmarking is used to measure, compare, and improve key processes.
		4. Individual staff members can determine their impact on goals and know their status against goals.
		5. Suppliers can measure how well they perform against standards.
		6. Processes for setting measures and comparing performances are valid and reliable.
		7. Documentation describes the standards and measures, the rationale, how they were determined, and how they are applied.
		8. The documentation describes how data are treated, interpreted, and applied.
		9. The measures are used to evaluate staff, suppliers, products, services, processes, and the overall function performance.
		10. What is learned is documented and shared within the function.

Source: reprinted with permission of the International Board of Standards for Training, Performance, and Instruction (Barrington, Illinois). © IBSTPI, 1994.

FIGURE 5.10. Number of IBSTPI's Standards, Performance Statements, and Criteria

	Standards	Performance Statements	Criteria
Instructional Designer	16	66	186
Instructor	14	83	403
Training Manager	15	79	412

panding for every organization. This is particularly true for companies seeking either Baldrige or ISO recognition. In fact, it's virtually impossible to accomplish this goal without a major effort by the training function. Organizations throughout the country have changed dramatically. Profit margins are smaller and companies have downsized. They are faced with four very difficult situations:

- There are fewer employees. Those that remain must now do tasks once relegated to others.

- The nature of many jobs has changed. Widespread use of personal computers and less reliance on mainframes, for example, make it important for everyone in an organization to become computer literate.

- To gain market share internationally, companies must compete with ISO-qualified organizations at home and abroad.

- The basic nature of work has changed. Where once the emphasis was on marketing, production, and distribution, companies must now concentrate on quality and cost.

All of these situations have very large implications. Without adequate training, companies will not be able to compete.

FIGURE 5.11. Training Manager Standards

1. Assess organizational, departmental, and program needs.
2. Develop plans for the department and programs.
3. Link human performance to the effectiveness of the enterprise.
4. Apply instructional system design and development principles.
5. Assure the application of effective training principles.
6. Evaluate the instructional design, development, and delivery function.
7. Apply the principles of performance management to own staff.
8. Think critically when making decisions and solving problems.
9. Ensure actions are consistent with goals and objectives.
10. Adapt strategies and solutions given change.
11. Produce effective and efficient solutions.
12. Develop and sustain local relationships.
13. Provide leadership.
14. Use effective interpersonal communication techniques.
15. Communicate effectively orally and in writing.

FIGURE 5.12. Instructional Designer Standards

1. Determine that the project is appropriate for an instructional solution.
2. Conduct a needs assessment.
3. Assess relevant learner characteristics.
4. Analyze characteristics of the setting.
5. Perform job, task, and content analyses.
6. Write performance objectives.
7. Develop performance systems.
8. Sequence the performance objectives.
9. Specify the instructional strategy.
10. Design instructional materials.
11. Evaluate the instruction.
12. Design an instructional management system.
13. Plan and monitor instructional projects.
14. Communicate effectively visually, orally, and in writing.
15. Interact effectively with others.
16. Promote the use of instructional design principles.

FIGURE 5.13. Instructor and Trainer Standards

1. Analyze course materials and learner information.
2. Assure preparation of the instructional site.
3. Establish and maintain instructor credibility.
4. Manage the learning environment.
5. Demonstrate effective communication skills.
6. Demonstrate effective presentation skills.
7. Demonstrate effective questioning skills and techniques.
8. Respond appropriately to learners' needs for clarification or feedback.
9. Provide positive reinforcement and motivational incentives.
10. Use instructional methods appropriately.
11. Use media effectively.
12. Evaluate learner performance.
13. Evaluate delivery of instruction.
14. Report evaluation information.

At the same time, the old way of training on-the-job (coupled with presentations in classroom settings) has become too cumbersome and unreliable. Standards can help training accomplish its mandate with far greater efficiency and effectiveness than ever before.

Compliance with Baldrige, ISO, or TQM requirements will be very difficult for an organization if its training function can't meet the requirements internally. Training can't reasonably expect to deliver quality unless it can exemplify quality.

Check Your Understanding

The following questions will give you an idea of how prepared your function is and where you might want to improve. Use a check mark to show your function satisfies a criterion, an x to show it doesn't, or a zero if you aren't sure whether it does or not.

____ 1. Can your customers discuss the learning and performance improvement issues facing their organization?
____ 2. Can training staff discuss the organization's important business issues, strategies, and key initiatives?
____ 3. Can your customers discuss how you measure the effectiveness of your learning and performance improvement efforts?
____ 4. Can your customers discuss your vision of quality for the learning and performance improvement function?
____ 5. Can your staff discuss your vision of quality for the learning and performance improvement function?
____ 6. Can your staff discuss the standards and processes used for identifying, designing, developing, and delivering learning and performance improvement efforts?
____ 7. Does your staff act independently on behalf of internal customers?
____ 8. Is your training function's vision for quality consistent with and supportive of the vision and key initiatives of your organization?
____ 9. Does the training function's vision serve as a guide in making decisions about the design, development, delivery, and evaluation of learning and performance improvement efforts?
____ 10. Does your staff initiate action to prevent vendor programs or services from deviating from standards?
____ 11. Does your staff initiate, recommend, or offer new standards, and methods of compliance?
____ 12. Does your function conduct itself in a manner consistent with the performance values it espouses and with respect for its individual members?
____ 13. Are your standards for instructional programs documented and followed?
____ 14. Are your standards for selecting vendors and consultants documented and followed?
____ 15. Are your standards for instructors and facilitators documented and followed?
____ 16. Can you link the results of your efforts to the effectiveness of your organization?
____ 17. Do your practices match the core values of your organization?
____ 18. Do you have processes in place for training your staff in what your standards are and how to apply them?
____ 19. Do you have processes in place to verify how vendor products and services meet your standards?
____ 20. Do you have processes in place to verify customer needs?
____ 21. Are your processes documented?
____ 22. Is your staff trained in how to use these processes?

_____ 23. Do you have a process in place for assessing the qualifications of your staff?

_____ 24. Can you produce an audit trail showing adherence to your standards and processes?

_____ 25. Do you have in place processes that allow you to determine what the cycle time was for design, development, delivery, and evaluation of learning and performance improvement efforts?

_____ 26. Is your process for assessing your organization's needs documented?

_____ 27. Do you know the alternatives available to your customers for learning and performance improvement services?

_____ 28. How do your services compare to the alternatives?

_____ 29. How do your costs compare to the alternatives?

_____ 30. Do you know the cost competitive implications of being more or less responsive to your customers?

_____ 31. Do you know what your customers value about your service?

_____ 32. Do you know your costs in dollars, time, and resource commitment?

_____ 33. Do you know what factors contribute to costs?

_____ 34. Do you join your customers in developing change strategies for your organization?

_____ 35. Do you initiate the need for change strategies based on what you learn about the issues facing your customers?

_____ 36. Do you engage in conversations about change and change strategies with your customers and staff?

_____ 37. Do you engage in conversations about the need for change strategies within your own function?

_____ 38. Does your staff initiate conversations about the need for change, how to support change, and how to develop change strategies?

_____ 39. Do you share with your customers the responsibility for examining the necessity and the trade-offs of changing?

_____ 40. Do you manage facilities or other physical resources associated with the deployment of learning and performance improvement activities?

_____ 41. Do you have criteria for evaluating the cost effective use of those resources?

_____ 42. Do you evaluate how well those resources are satisfying your customers?

_____ 43. Do you compare the costs of using alternative resources?

_____ 44. Do you have plans in place to optimize the use of your resources?

_____ 45. Do your customers prefer to use your resources?

_____ 46. Does your staff speak well of your resources?

What to Do Next

1. Meet with your staff or assign a team to discuss their answers to these questions. Have them discuss how standards might help them communicate what they do and what value they bring. Share your ideas from the notes you've taken. Have them decide what standards they would like to develop and apply.

2. Meet with your staff or assign a team to review the ISO training standards. Have them discuss:
   — How well the function meets those standards.
   — What documentation they would show an examiner.
   — The benefits of working toward these standards.

3. Meet with your staff or assign a team to review the Baldrige criteria for customer satisfaction and quality results. Have them develop a set of standards for how training will:
   — Involve its customers.
   — Determine what products and services to offer.
   — Assess the performance of staff, products, and services.
   — Ensure standards remain useful.

4. Meet with your staff or assign a team to review IBSTPI's standards. Have them decide which standards:
   — Will improve your department's performance.
   — To adopt.
   — To modify.

5. Meet with your staff or assign a team to benchmark with other training organizations to see how they use standards.

6
MEASUREMENT AND EVALUATION

The peregrine falcons living in Chicago's skyscrapers provide the basis for an interesting discussion about measurement and evaluation. What factors about the falcons should be measured? A biologist might measure the size of a bird or how fast it can fly. A naturalist would be interested in how well it is doing in its new environment and whether it can reproduce satisfactorily. A social psychologist might want to determine its affect on the people living in Chicago. A physicist might study how it flies and rides thermals.

What about you? As a training and development professional, what would you want to evaluate about the peregrine falcons in Chicago? What does our discipline indicate? What questions might you ask? Think about it for a moment before continuing.

The peregrines were taken to Chicago for two purposes; to provide them with a congenial new environment, and to get some control over the pigeon population in the city. To address the first concern, we might want to determine whether the birds are mating in the city and, if so, how many chicks are there and are they healthy? Will the chicks stay in the city, will they be content to remain in the new environment? What about the pigeon population? Is it growing or falling? What will happen when there aren't enough pigeons for the falcons? Will they look for new prey? If so, what will they pick? Something we would like to see them choose, such as rats, or will they prefer doves and robins?

All of these questions should be answered. Our quality orientation demands the answers be determined by measurement, not someone's opinion, rigorous fact based answers so the city can take intelligent action if needed.

This chapter builds on the premise that the quality movement, among other things, is driving training toward better evaluation of its processes, products, and services. It is about:

- What to measure and why.
- The measures used by training.
- How to design evaluations.

WHAT TO MEASURE AND WHY

Four Levels of Evaluation

Traditionally, training considers four levels of evaluation.[1] The four levels can apply to any evaluation. The four are:

1. Liking or acceptance.
2. Learning or acquisition of new skills and knowledge.
3. Application or transference to the job.
4. Accomplishment or satisfaction of a customer, business, or regulatory need.

All four levels have value when used properly. How and why each level is used is described in greater detail in the following sections.

Level 1. This level answers the question, "Did the learners like it?" Evaluation at this level measures learner acceptance of the experience. The so-called "smile test" is useful for finding out how learners responded to the experience. It provides information about the adequacy and comprehensiveness of the training as well as attitudes about the environment, people involved, and relative importance of the content to the individual respondent. Evaluation standards normally specify how attitudes are to be surveyed.

---

[1] D. Kirkpatrick, *Evaluating Training Programs* (Alexandria VA: American Society for Training and Development, 1975).

Level 2. This answers the question, "Did they learn?" Evaluation at this level is a test of a learner's grasp of the knowledge and skills offered during training. Final examinations (or end-of-course tests) or check lists of criteria are used to probe the amount and depth of knowledge and skills learners have acquired. Evaluation standards usually refer to how learning is assessed (e.g., performance test, criterion-based) and how a test is administered.

Level 3. Evaluation at this level answers the question, "Did they apply what they learned on the job?" This level measures how well learners transfer new skills to the job or other appropriate venues away from the training environment. The usual way to find out if transfer has occurred is to conduct an audit downstream from the training event. The performance audit can be an important useful tool for this activity.[2] Evaluation standards include criteria for observation guides and ways to measure change in job performance.

Level 4. This level answers the question, "Did it make any difference?" This type of evaluation assesses the organizational benefits that accrue from the training. This sort of evaluation is sometimes made difficult by the conflicting interests of stakeholders. One method is to select, before training takes place, artifacts or outputs that should change in predictable ways if the training is effective. When training is completed, an audit should be led to compare the new, current state of things with what was before the training happened. Standards for this level of evaluation are rare, but one can assume they would indicate how much improvement should take place and, in general terms, the nature of that change.

To summarize, trainers can measure to see if learners value the experience (level 1), learners achieve the learning goals (level 2), learners apply what they learned (level 3), and the right business need was met (level 4).

---

[2] O. Westgaard, *The Competent Manager's Handbook for Measuring Unit Productivity* (Chicago: Hale Associates, 1988).

Training typically evaluates level 1. For some programs, learning is evaluated (level 2). However, ISO certification is primarily interested in levels 2 and 3. ISO's standards call for people being qualified to do the job. This means they **can** (they know what to do and how to do it) and **do** (they apply what they know to the job). Baldrige measurement of quality results and customer satisfaction calls for levels 3 and 4. Baldrige examiners want to know that not only are people qualified, but they produce products and services that meet standards and are valued by customers. They do what's right, they do it well, and they do what matters.

Evaluations done during the training are sometimes called *formative evaluations*. They tell what, if any, corrective action needs to be taken. Evaluations done at the end of the training are called *summative evaluations*. They tell what people felt about the experience and if learning occurred. Level 1 evaluation can occur during and at the end of the instruction. It is both formative and summative. Level 2 evaluation occurs at the beginning of the training, at discrete points during the training, and at the end. Like level 1, level 2 is both formative and summative.

Evaluations done after a training event are called audits. They tell if the need was satisfied. Level 3 and 4 evaluations are summative. They are done after the training. Enough time has to pass to find out whether or not the new knowledge and skills are really being applied and the consequences of their application. A picture of when these different evaluations occur might look like Figure 6.1, which represents the delivery of one training event. That could be a workshop, a

FIGURE 6.1. Timing of Evaluations

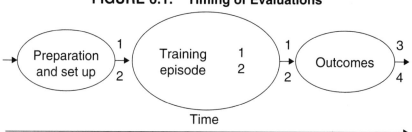

self-study program, or a module delivered by computer-based training (CBT). The numbers show when each level of evaluation occurs in training over a period of time.

Uses of the Four Levels

Level 1 evaluation is used to sample opinions. How do learners feel about a course they took? Most surveys are aiming for level 1 responses. Market research depends heavily on this sort of measurement as do some needs assessment efforts. For training, the learner can be asked to comment about the:

- *Content.* How relevant it is.
- *Exercises.* How useful they are, how well they build confidence and assure the learner that he or she is learning.
- *Organization.* How well the program is structured and if it makes sense.
- *Materials.* How easy they are to use.
- *Environment.* How comfortable the room is and whether or not light, noise, temperature, or layout support learning.
- *Trainer.* How well the trainer presented the topic, involved the learners, and managed the group.

Figure 6.2 is a sample of a survey used to evaluate learners' liking and acceptance of a training program.

In terms of usefulness, level 1 measurement is valuable because it can sample intangibles such as customer acceptance. It tells a trainer or course administrator what a learner likes and dislikes about a program. It is extremely important to measure customer satisfaction with what is done, how it is done, and the relationship it has with their own goals and aspirations. But it is not of much use in determining whether criteria are met. Getting someone's opinion about a standard

FIGURE 6.2. Session Evaluation Form

Session Title: _____ Date: _____

Trainer(s): _____

I. Session Content:

   A. How relevant was the session content to your current work?

Very Relevant	Somewhat Relevant	No Opinion	Somewhat Irrelevant	Very Irrelevant
5	4	3	2	1

   Comments: _____

   B. Was the content sufficient to meet the objectives for the session?

Very Sufficient	Somewhat Sufficient	No Opinion	Somewhat Insufficient	Very Insufficient
5	4	3	2	1

   Comments: _____

II. Session Organization:

   A. Were the topics within this session arranged logically?

Very Logically	Somewhat Logically	No Opinion	Somewhat Illogically	Very Illogically
5	4	3	2	1

   Comments: _____

   B. Was the time allowed for each topic adequate?

Very Adequate	Somewhat Adequate	No Opinion	Somewhat Inadequate	Very Inadequate
5	4	3	2	1

   Comments: _____

does not tell us whether they have accomplished it. Instead, level 2 evaluation is needed.

Level 2 evaluation is frequently referred to as the pre- and post-tests. The reason for doing a pretest is to confirm what the learners know (and do not know) before the training. A trainer wants to prove that what they learned was a result of completing the training and not from some other source. The important thing is level 2 requires some kind of test. The type of test depends on what is being tested (i.e., knowledge or skill).

Performance tests and checklists are instruments used to test skills. They are also useful for deciding the amount of skill a person has. They are used to separate unskilled from competent and competent from exemplary. They discriminate very well. Because of this, they are useful in measuring the ability to accomplish standards. Figure 6.3 is a sample of a pretest used for a workshop on evaluation.

Evaluation can determine whether workers are capable of producing quality products. It can reveal whether an insurance agent has the skill to close on a contract. However, level 2 evaluation does not establish whether people apply what they learned or accomplish what they are supposed to.

Level 3 evaluation takes up where level 2 leaves off. It can be used to learn if people use their skills to achieve the quality standards set. Level 3 is also done to find out if training is an efficient and effective way to transfer skills and ensures learners are following procedures designed to protect them, other people, or physical and financial assets. In this case, training is frequently done in response to regulatory requirements. Two examples are safety and cash handling procedures.

Baldrige and ISO examiners who go into an organization to observe people doing their jobs are measuring at level 3. They are auditing work in progress. They also audit the completion of tasks by studying documents and audit trails established for that purpose.

Level 3 evaluation can be done a number of ways. For example, it can be done by:

- Watching a person at task and comparing what is observed against a set of criteria on a checklist.
- Asking (through surveys and interviews) the person, coworkers, and supervisors to report changes in behavior and in the quality or quantity of work produced.
- Examining the person's work (what is actually produced) and comparing it to standards.

FIGURE 6.3. Pretest

1. What is one purpose of using a pretest?
   a. To provide a practice test-taking opportunity.
   b. To decide if teaching the topic is necessary.
   c. To dotormine if learners have come to class prepared.
   d. To ascertain if learner knowledge has increased.
   e. Don't know or not sure.
2. You are an internal consultant for a medium-sized corporation. You have been asked to discover if a gap exists between the skills a group of employees now have and will need in the future to perform a new task. What process will you use?
   a. Analysis.
   b. Assessment.
   c. Evaluation.
   d. Measurement.
   e. Don't know or not sure.
3. You are an internal consultant for a medium-sized corporation. You have been asked to find out why the productivity of a group of employees has risen dramatically in the past six months. What process will you use?
   a. Analysis.
   b. Assessment.
   c. Evaluation.
   d. Measurement.
   e. Don't know or not sure.
4. What evaluation strategy is most appropriate for discovering whether learners have been able to apply newly learned skills to their work?
   a. Conduct a performance audit.
   b. Assess needs at the micro level.
   c. Analyze the results of the post test.
   d. Survey supervisory personnel.
   e. Don't know or not sure.

Answers: 1.b, 2.b, 3.d, 4.c

The issues related to doing level 3 evaluation are:

- What evidence will prove on-the-job behavior has improved as a result of training?

- When will the training function secure the evidence?
- How will it secure the evidence?
- What will it do with the evidence?
- Who will it give the evidence to?

Figure 6.4 is a sample of an observation form used to evaluate whether trainers follow the procedures they learned in a course on how to be an effective trainer.

Although transference back to the job is important, level 3 still does not tell whether training actually met the business need. For that, level 4 evaluation is needed.

Level 4 evaluation steps away from the individual (or specified group) and looks to the organization and to customers. Level 4 measures, therefore, are critical at some point in the measurement of quality standards because they are the only level based on the customers' point of view. They consider the products and services of the organization in the light of the customers' wants and needs. The evaluation is done to correlate the training to the satisfaction of the business need. Figure 6.5 lists the types of business needs training is expected to affect.

For example, a customer accounts receivable department has aging receivables of 90 days or more that average $5 million. A training course is given on collections procedures. A level 4 evaluation would be to compare the amount of aging receivables before and after training.

Another example is reducing the length of the sales cycle. The sales cycle is the length of time from the initial customer contact to securing a contract. If training can shorten the cycle, customers can generate revenue earlier. A level 4 evaluation would be done to compare the length of the sales cycle before and after training. Level 4 evaluation is done by performing the following:

- State the business need for the training before it is offered.

FIGURE 6.4. Sample Checklist

Competency 3: Establish and maintain instructor credibility.
Performance 1: Demonstrate acceptable personal conduct.

Criteria	Basic	Intermediate	Advanced	Yes	No	N/A	Comments
The instructor:							
1. Wears clothing and is groomed appropriately for the audience and the situation.							
2. Demonstrates good posture and carriage.							
3. Admits without bias errors made and areas of ignorance.							
4. Promotes transfer of skills, knowledge, or attitudes from the instructional setting back to real-life.							
5. Obtains and transmits information requested by learners.							
6. Invites and positively accepts feedback regarding performance as an instructor.							
7. Where appropriate, accepts responsibility for, without blaming or belittling others, the learning materials, management, etc.							
8. Demonstrates behavior in accordance with a code of ethics comparable in intent to that published by the International Board of Standards for Training, Performance and Instruction.							

FIGURE 6.5. Business Needs Affected by Training

Financial. • Profitability. • Cost containment. Performance. • Cycle time. • Resource time. • Quality. • Customer satisfaction. • Productivity. Market. • Share. • Penetration.	Compliance. • Corrective. • Avoidance. Image. • Association. • Disassociation. Personnel. • Recruit/retain. • Retool/develop.

- Select measures that are meaningful to the client.
- Make it explicit how they link to business need.
- Confirm the data are accessible.
- Identify any opportunity for corroborating evidence.

THE MEASURES USED BY TRAINING

What about evaluating the training function's effectiveness? This is most often measured by level 1 (liking) and level 2 (learning). However, there are two other measures used by training: training's own productivity and return on investment.

Training's Productivity Measures

Training typically measures its own output. Those measures include:

- The number of student hours.

- The number of times a course was offered.
- Enrollment by course and by instructor.
- The percentage of time training rooms are in use.
- How many times or hours a video or CBT program was in use.

In the past these measures have been adequate. They were considered productivity measures. The more or the higher the number or ratios the better.

Today, however, the questions are, "What should be measured?" and "Against what standards?" The quality movement has changed the answers to both. What the quality movement has taught is that the standards are set by the wrong people and the process is closed rather than open to changing preferences. Customers may no longer value the same measures. Regulators, for example, were satisfied with output measures (i.e., the number of people trained in a procedure and the length of the training). Regulators are now asking for proof that trainees learned and are complying with what was taught. They want levels 2 and 3 evaluation. Customers and the organization's leaders are now asking for level 4 evaluation. They want proof the investment was worth it and the need was satisfied.

Return on Investment

Another measure used by training is return on investment (ROI). Figure 6.6 shows what happens before and after training is offered and how that information is used to determine ROI.

The investment is all the costs incurred to get the customer's criteria and to design, develop, produce, and deliver training. Investment measures are around training processes: cycle time, consistency (i.e., reduced variability), continuous improvement, and resource use.

FIGURE 6.6. Calculating ROI

Investment				Return
Inputs	Processes	Products	Outputs	Outcomes
Customer expectations Business needs Regulations	Task analysis Design and develop training	Course materials	Course delivery	Customer satisfaction Business results

Management has rediscovered Deming's and Juran's ideas on:

- The proper use of process controls.
- How to define processes.
- How to track and measure deviation from standard.

Training is not exempt. It must evaluate the quality of its critical inputs. For example, a training course is built on inputs from content (i.e., subject matter) experts: The quality of the evaluation must be evaluated before it moves through the development process into the course design phase. How many courses are redesigned because the inputs were never qualified? Think of impact on quality and price if training chooses to set standards for and measures every part of its processes.

Beyond providing learning events for the organization, training conducts needs assessment, job and task analysis, and formative and summative evaluation of its impact. All these activities rely on measurement. Standards, however clear, are useless unless they are used. Their use requires measurement—data gathering—and evaluation of the results. The adequacy of any measurement effort depends on the criteria.

The measures of return are stakeholder satisfaction, regulatory compliance, return on investment ratios, and the effect on business. Therefore, to get the return requires level 4 evaluation. Level 4 evaluation can be done without calculating the ROI. However, an ROI cannot be done without a level 4 evaluation. A level 4 evaluation cannot be done without adequate description of the customer's criteria for satisfaction or the business need.

Training figures out the wants and needs of customers whenever it performs a needs assessment. Therefore, if training is seen as the supplier of credible, accurate, and reliable information on internal customers, it is in position to take on a leadership role in measurement and evaluation of those customers' wants. It is the customers' criteria that becomes the measures for level 4 evaluation.

HOW TO DESIGN EVALUATIONS

Formal Investigations

Often it becomes important (because of the extent, criticality, or controversial nature of the subject) to use a formal investigation process. The process has five steps:

1. *Set the stage.* State the problem or condition that requires investigation, do a preliminary cost/benefit analysis, and develop hypotheses.

2. *Select a method and design the investigation.* Decide how the research will be done.

3. *Implement.* Do the experiment, conduct the survey or whatever is planned.

4. *Collect and interpret data.* Assemble the data in some kind of useful form and then analyze it using a formal, accepted procedure.

5. *Develop recommendations and conclusions.* Finally, evaluate and develop reasoned arguments about what the results mean and why.

The following four examples provide a basis for designing custom evaluation efforts.

Example: Merchandising Training and Transactions. A retailer wanted to know if training in merchandising practices would be worth the effort. It wanted to know if (and if so, by how much) training in merchandising principles would increase sales transactions. It selected 45 retail outlets. The process was to record the number of transactions before and after training and compare the results with transactions in outlets that did not receive training (a control group).

The number of transactions were carefully recorded in each of the 45 retail outlets for 120 days. The merchandising materials in each store were the same and used the same way. This data was recorded in tables so it would be easy to use later. After 120 days, the managers in 35 of the outlets were trained in how to use merchandising materials for the next promotion. Managers in the other 10 stores were not trained. Again, transactions were carefully recorded for 120 days just as before.

The data from the before and after results developed into two important sets of information. First, there was a slight increase in transactions that could be credited to the second promotion. Second, no significant difference was found that could be credited to training. Therefore, it was concluded training was not justified.

Management Training and Sales Volume. The Walgreen Company is one of the largest drug store chains in the world. Their pharmacies are located in neighborhoods that vary immensely in environmental conditions. The differences are climatic, ethnic, economic, and political. Despite these differences the company strives to offer top quality service at every location. Corporate training at the Walgreen Company was included in this effort more than 10 years ago and has been able to perform several studies that are both interesting and informative. The management team of Jim Shultz and Anne Laures has directed these efforts.[3]

One of the first studies they did was to try to find out if training could help improve the performance (i.e., amount of merchandise handled) in particular stores. They chose 10 stores from the same geographic area serving approximately the same kinds of customers. Five of the stores were doing well; their sales volumes were at or above the national aver-

---

[3] A. Laures and J. Macygin (Unpublished interview, Deerfield, IL: Walgreen, 1994).

age. Five were doing poorly. The experiment was designed to train all 10 managers in specific ways using the same training techniques and materials for all of them. They trained all 10 simultaneously at the same facility. The hypothesis was that, after training, the average volume of sales for the 10 stores would rise, the poorer performing stores would increase sales significantly more than the others, and the variance between good performers and poor performers would lessen. Average sales did increase significantly, but the variance wasn't affected much. The training was justified but the problem of poor performers persisted.

This study was important for two reasons. First, by the way it was designed, all the stores were able to take advantage of the intervention. There was no need for a control group. Second, because the investigators had hard data about sales they could use statistical analysis on the results. The results they reported were supported by inferential analysis and, therefore, much easier to sell to corporate management.

Tuition Reimbursement and Turnover. Walgreen also did a study of a non-training intervention. As with many companies, the ability to lure and keep competent employees is paramount at Walgreens. They are particularly interested in keeping part-time store employees (i.e., those who work less than 30 hours per week) on the payroll. These jobs do not pay high wages. Consequently, turnover can be very high. Walgreens decided to begin a program to help its employees attain college degrees. It was a tuition reimbursement program with continuing public relations efforts to recruit college-bound workers and remind current employees to take advantage of the program. The theory was that if people used the program, they would remain part-time employees while attending a local college or university. Given the number of part-time employees Walgreen hires nationwide, a small decrease in turnover would result in millions of dollars in savings to the company.

According to an internal report prepared by the training and development department at Walgreen, "In 1990, our part-time turnover was running 192% [annual basis], versus our full-time turnover of 5.9%. . . . The high part-time turnover was not giving much pay back on our training investment; the part-timers constituted almost half our work force, and they were not around long enough to really become competent and proficient in their jobs. . . . "[4] This data would be useful later when results of the study were analyzed. Direct comparisons could be made.

The goals of the program were to:

- Find ways to help employees develop skills, so they could look forward to higher levels of responsibility and performance.
- Build good relations within the communities served.

The tuition reimbursement program was unique. "We designed our program to be less restrictive than most tuition programs:

- No minimum grade requirement.
- No job-relatedness requirement.
- Immediate enrollment upon hire.
- Ability to use funds two years after leaving company.

Our only requirements were averaging 15 hours/week, working 60 days before drawing any funds, and ensuring the learning institution was licensed."[5]

The department picked four retail districts that were very similar in size, demographics, and revenue for the study. The program was started in two of the districts while the other two were used as controls. The stores in the experimental districts were compared with the stores in the control districts to see if the program had an impact on part-time turnover.

---

[4] J. Macygin, *Report on Impact of the WISE Program* (Deerfield IL: Walgreen).
[5] Macygin.

Measurement and Evaluation

They compared the stores by "comparing the average difference in profitability between high turnover stores and low turnover stores, and applying that difference to the test stores" and by "dividing the total amount of dollars paid in unemployment compensation by the number of people turned ... to get a cost per turn."[6]

"However, we found no significant turnover reduction in the test districts' compared to the control districts':

- Turnover among full-time employees.
- Turnover among part-time employees.
- Turnover among employees 18 and younger.
- Average turns per month.

When we compared the nine months preceding the tuition program with the nine months following the program introduction, turnover was in fact lower in all four districts; but the tuition program (WISE) did not explain the lower turnover in any of the analyses we ran."[7]

Walgreen's training and development also used a survey for this study. "Prior to program implementation, we randomly selected 986 employees from the four districts; store managers evaluated those employees on customer service, cash handling, merchandising/stocking, housekeeping, reliability, attitude, and trainability. ... A year later we surveyed again. ... We saw no significant performance improvement; so WISE did not seem to motivate these employees to higher levels of performance."[8]

They also did not see any difference in performance evaluations. They did, however, discover that workers who were considered good before the program began tended to take advantage of it. Only 8.4% of the employees hired during the

---
[6] Macygin.
[7] Macygin.
[8] Macygin.

period knew about the tuition program before applying. Of those, less than 30% said it affected their decision to apply. Consequently, Walgreen is phasing out the tuition reimbursement program.

Training for Pharmacy Techs. The final example from Walgreen's concerns a training program. Walgreen has a class of employees called *pharmacy techs*. These people provide assistance to registered pharmacists. Currently there are two methods for training them. One is a formal program conducted by the training and development department consisting of 20 hours of classroom instruction followed by 20 hours of supervised practice. The second method is on-the-job-training (OJT) conducted by the pharmacist and other pharmacy techs.

The purpose of the study was to discover any differences between the results of the two types of training and what impact those differences made on return on investment. The primary data gathering instrument was a survey prepared specifically for pharmacists. Analysis of the data showed significant differences in several areas. For example:

- 29% of the formally trained people asked the pharmacist for help while 42% of the OJT group did so.

- 66% of the formally trained people were considered exemplary at providing window service, compared to 47% of the OJT group.

- 88% of the formally trained people could direct customers where in the store to find an over-the-counter drug they asked for, while only 56% of the OJT group could do the same.

- The formally trained pharmacy techs generated about $1,210 more in sales than the OJT group in their first month of service.

Most likely, there will be a strong recommendation for store managers to take advantage of the formal training program for pharmacy techs.

Informal Investigations

Formal studies take time and money. They often require expertise beyond what a small training function can muster. Is there an alternative? Yes, there are dozens. Informal investigations are taking place constantly. For the most part though, they're unplanned and poorly documented. For example, one of the most effective observation techniques is a simple phone log: who called, when, for what reason, and the responses they received can provide a great deal of pertinent information. Yet, the activity costs little or nothing and does not take much extra time.

People in the training function are often in conversation or meetings with people from other functions. Simply taking notes during these opportunities documents data for later assimilation and analysis. With a bit of planning the training function can take advantage of these and other opportunities to lay the foundation for a more formal effort.

There is much more that can be done. For example, think about the training director who wanted very much to know what effect an old course was having on participants; how well they were transferring what they learned in the classroom to actual practice in the field. Since it was a well-established course, however, that had been accepted by management as appropriate and valuable for years, she couldn't get them to sponsor a formal study. She sent one of her people to talk to a recent trainee in a nearby facility and to watch the learner try to accomplish a task using the new skill set. An hour later she had much better information. She used it to get management to let her survey supervisors companywide to see how many had the same kinds of trouble.

The tools used for formal studies are also valuable with informal investigations. Training may not be able to use them in the same way or with the same amount of accuracy, but they can still be appropriate and provide important information.

Measured Results

A measured result is one that can be stated through "hard" data. Hard data can be analyzed systematically and thoroughly. Normally, hard data can be represented with numbers or other symbols that can be manipulated mathematically.

It's important that whoever is trying to establish compliance with standards such as ISO do so through the use of measured results. Otherwise, the measurement suffers from lack of precision, rigor, and, therefore, credibility. Standards, themselves, should be presented so that they will require the use of measured results. This is often difficult, but it is provident. Some examples are measures to evaluate the efficiency with which products and services are developed and provided (e.g., cycle time, response time, defects, rejects, rework, or ratio of energy required to output generated). It would be simple to say something like, "Rejects shall be kept to a minimum." One person's opinion of what a minimum is will probably not match another's. The standard writer should say something like, "The number of rejects shall not exceed 12% of the number meeting the standard." Coming up with such a percentage requires a great deal more effort than simply saying rejects are bad. But the effort is usually well worth it.

One alternative to measured results is to compare the results of one operation with another. If plant A turns out 1.2 million candy bars per week, plant B will be expected to do as well or better. Given that they are the same kind of candy bars and both plants have the same equipment, the comparison is still a poor measure because it does not establish goodness. What if plant A produced 5 million, or 0.5 million? Comparing B with them becomes ludicrous. "Keeping up with the Jones's," has never been a good way to establish standards. When tempted, don't. Instead, use measured results.

Measured results are central to the thinking of both Deming and Juran. One of Deming's key indicators of quality is variability of product or service. Quality, in his opinion, is partially achieved through reduction of variability. Customers want to know what they are buying. For example, a company ordered a training event for a work group. Employees said it was the best training they every had. So the next time the customer needed that training, the company ordered it again. This time the work group said it was terrible. That customer may not go back for more training.

So how could training get measured results to reflect the customer's new attitude about training? A level 1 measure is needed to evaluate through the use of hard data. For example, the work group could be asked:

What is your opinion about the relative quality of the training?

1. Great.
2. Very good.
3. Average.
4. Very Poor.
5. Total waste.

Or training could conduct a structured interview with a sample of the work groups, or do a structured observation of on-the-job performance. Whatever evaluation training does (it could do all four levels) should be designed to yield hard data. Then the results will clearly show whether the event met quality standards.

Documentation

Good measurement practices require documentation. It makes sense but it is often overlooked or given less attention than it should have. Documentation provides access to data and information. Without it, people who should know certain things won't, or won't know all they should as precisely as they should. That sounds like double talk. For example, an instructor has just finished a session for a group. At the end, rather than ask them to fill out evaluation forms (i.e., smiles sheets), she asks them to raise their hands if they felt the session went well. Twelve of the fifteen raise their hands. The instructor asks each of the others what their opinions are. One says the session was too long, the second says it was hard for him to hear the instructor, the other one is not very specific. What will the instructor say to her boss when asked how the session went?

If the instructor had written down how many raised their hands and the comments of the dissenters, she would have had much better information for her boss. Better yet, if she had used a prepared instrument to get the data, her information would have been even more pertinent and credible. This simple example points directly to the need for adequate comprehensive documentation procedures and practices for the training function.

This book is not the arena for a long discussion of what documentation is and why it's important. So we have to be content in saying it is important and it's important for every aspect of the measurement effort. Training should document the design, development, implementation, and evaluation tactics, as well as customer responses, carefully and as completely as possible. There can never be too much documentation, but there can be too little.

SUMMARY

Evaluation can be a challenge. The opportunities to measure and evaluate the training function's performance include:

- Accumulating hard data from needs assessments.

- Analyzing trends and their impact on future productivity.

Measurement and Evaluation 169

- Diagnosing gaps in the development of critical competencies.

- Certifying that customers have learned what they set out to learn.

- Auditing the impact of training outcomes on the business processes.

In turn, training must decide what to measure, when to measure, how to measure, and what methods or processes to use to get valid, reliable data. Skill in measurement and evaluation is becoming more valuable to organizations every day, particularly those implementing quality programs.

Check Your Understanding

This exercise is a bit different from the ones you've found in the other chapters. It is intended to confirm your understanding of your training organization's evaluation capability. In the column that asks if you have the tool, describe the tool you have that is best for the task. If you can honestly check all the yes spaces in the fourth column, you have a uniquely equipped and qualified training function.

Do you have access to this tool?	Yes	Can you use it effectively?	Yes
A procedure for conducting observations of people performing hard skills tasks.			
A procedure for observing people in meetings and ordinary conversations.			
A procedure for conducting open ended, investigative interviews.			

Do you have access to this tool?	Yes	Can you use it effectively?	Yes
A format and process for conducting structured interviews.			
A procedure for developing random samples.			
A format and process for developing assessment check lists.			
The rules and a process for developing surveys (either phone or paper and pencil).			
The rules and a process for developing a performance test.			
Inferential statistical methods that can be used to analyze test or survey data.			
The ability to develop graphic data displays.			
People who are experienced in developing interventions other than training.			

Next Steps

1. Select those programs and services you want to evaluate and at what level you want to evaluate them. Identify what information you still need to do the evaluation you want. Identify where, from whom, how, and when you will get the information.

2. Meet with your customers to develop more detailed descriptions of what they:
   A. Want to accomplish.
   B. Now use as evidence that a need exists.
   C. Will use as evidence that improvement occurred.
   D. How they might help you get the evidence.

3. As new requests for training and other services come to you, build into the process ways to get the information and measures you will need to do evaluations.

4. Meet with your training team and customers to discuss the three examples of how Walgreen evaluates its interventions. Discuss how you might use these as models for your training department. Discuss how you might achieve the same type of relationship with your customers to allow you to do similar work.

5. Build a plan to better measure your department's efforts.

6. Learn more about evaluation. Three references you can use to help you design processes tools are:
   — Robinson, D.G., and Robinson, J.C. *Training for Impact: How to Link Training to Business Needs and Measure the Results.* San Francisco: Jossey-Bass, 1989.
   — Rossett, A. *Training Needs Assessment.* Educational Technology Publications, 1987.
   — Westgaard, O. *Good Fair Tests: Test Design and Implementation.* HRD Press, 1994.

INDEX

ABB (company), 124
Administration, 85, 109–13
Advertising agencies, 18
Alliances, 16
Allstate Business Insurance, 17, 43–44, 45
Americans with Disabilities Act, 111
Amoco Information Technology Training (ITT) group, 41–42, 68–73, 94
  business processes, 71–72
  people, 73
  rewards, 72–73
  strategic direction, 69–71
  structure, 71
Amway (company), 123
Arthur Andersen & Co., 19
AT&T, 17
Attitudes, 42–44
Audit, 107, 148
Availability, 66

Baldrige Award. *See* Malcolm Baldrige National Quality Award
Behavior, 42–44
Benton Harbor community college, 19
Budget, 78–79
Burlington Northern Railroad, 19
Business case, 3–4
Business plan
  budget, 78–79
  definition of, 73–74
  elements of, 74–79
  executive summary, 75
  management team, 77
  marketing, 76
  operations, 77

Business plan (*continued*)
  organization, 76
  parameters, 75–76
  risks, 77–78
  schedule, 77
Business pressures, 20
Business processes, 71–72

Cadre, 32
CBT. *See* Computer based training
Change, 135
Commitment, 130, 133
Competence, 10–11, 24, 48
Competition, 18–20, 55–60, 79
Computer based training (CBT), 123
Contract services, 24
Cost, 66
Course maintenance, 109–10
Cultures, 43
Customer
  identification, 57–60, 62, 64
  relationships, 135
  requirements, 39, 60–62
  satisfaction, 2
Cycle time, 99–106

Deming, W. Edwards, 156, 166
Discontent, 54–55
Diversity, 12–13
Documentation
  evaluation, 167–68
  processes, 89–96
  services, 24
Downsizing, 15
Driving force, 38–39

Economic pressures, 18–20
Educational Testing Services (ETS), 123

Efficiency, 97–98
Employee competence, 10–11
Employer relationships, 16
Empowerment, 10
Environments, 24, 149
ETS. *See* Educational Testing Services
Evaluation, 145–70
  design of, 158–68
  documentation, 167–68
  formal investigations, 158–64
  four levels of, 146–58
    level 1, 146, 148–50
    level 2, 147–48, 150–51, 156
    level 3, 147–48, 151–53, 156
    level 4, 147–48, 153–55, 156, 158
  informal investigations, 164–65
  of leadership, 46–47
  measured results, 165–67
  of processes, 97–109
  standards, 125
Exeter, Thomas, 13
Expectations, 42

Federal Express (FedEx), 66, 123–24
Feedback, 86–87
Final analysis, 107
Florida Power and Light, 19
Ford Motor Company, 59–60
"Ford's Quality Education and Training Study," 59
Formal investigations, 158–64
Formal learning environments, 24
Formative evaluation, 107, 148
Formative feedback, 86
Front-end analysis, 124

Hard data, 165
Household International, 57

IBM Corporation, 19
IBSTPI. *See* International Board of Standards for Training, Performance, and Instruction

Image, 66
Imperatives, 43–44
Implementation, 125
Influence, 42–44
Informal investigations, 164–65
Initiation phase, 100–102
Input standards, 125–26
Instructional design, 24, 103
  standards, 138–39
Instructional program, 2
Instructors, 2, 24, 138–39, 149
Internal capabilities, 40
Internal discontent, 54–55
International Board of Standards for Training, Performance, and Instruction (IBSTPI), 25–26, 36, 50, 53, 56, 94, 123–24, 130–39
International Standards Organization (ISO), 4–12, 87, 124, 129–32, 136, 151, 165
Investigations
  formal, 158–64
  informal, 164–65
Investment, 20–22, 40–42, 156–58, 157–58
ISO. *See* International Standards Organization
ITT. *See* Amoco Information Technology Training (ITT) group

Jerry Company, 55
Job descriptions, 91
Job processes, 92
Johnson County Community College, 19
Juran, Joseph, 156, 166

Laures, Anne, 160
Leadership, 31–52
  in complex social settings, 46
  definition of, 32–33
  to enable learning, 47–48
  evaluation of, 46–47
  influence, 42–44

Leadership (*continued*)
  role, 37
  stewardship, 48–49
  system, 36–37
  use of power, 44–46
  values, 33–36
Learning, 47–48
  application of, 147, 151
  environments, 24
  grasp of knowledge, 147, 150–51
  learner acceptance, 146, 148–50

Malcolm Baldrige National Quality Award, 4–12, 33–34, 53, 56, 87, 136, 148, 151
Management team, 77
Management training, 159–60
Manipulator, 42
Marketing plan, 76
Materials, 149
McDonald's Corporation, 35, 39, 57
Measured results, 165–67
Measurement. *See* Evaluation
Measures of return, 158
Merchandising training, 159
Mission statement, 9, 37–42, 62–63

National Institute of Standards and Technology (NIST), 5
Needs analysis, 24
Needs assessment, 24
NIST. *See* National Institute of Standards and Technology

OJT. *See* On-the-job-training
On-the-job-training (OJT), 163-
Operations plan, 77
Opportunities, 67–68
Organization
  function, 71, 76
  program, 149
  and training, 147, 153
Outcome standards, 128
Output standards, 127–28
Outsourcing, 16, 66

PacBell, 19
Paradigm shift, 2
Participation, 8–9
Peregrine falcon, 1, 31, 53, 83, 119, 145
Performance, 91–92, 135
Pharmacy techs, 163–64
Post-test, 150–51
Power, 44–46
Preparation, 124–25
President's Quality Award, 23
Pre-test, 150–52
Pricing, 76
Processes
  administration, 109, 111–13
  business, 71–72
  conceptualizing and using, 106–13
  course maintenance, 109–10
  documenting, 89–96
  of evaluation, 97–109
  optimizing, 83–117
  redesign, 15–16
  staff development, 113–14
  standards, 126–27, 135
Process map, 94
Products, 24, 39–40, 65, 126–27
Profit, 40–42
Programs, 24
Purchasing officers, 35

Qualification, 124–25
Quality, 1–29
  Baldrige Award, 4–12, 33–34, 53, 56, 87, 136, 148, 151
  movement, 4–12
  President's Award, 23

Reality, 48
Regulators, 155–56
Relationship map, 104–5
Request for proposal (RFP), 65
Resource use, 104–6, 135
Return on investment (ROI), 20–22, 40–42, 156–58
Rewards, 72–73
RFP. *See* Request for proposal

Risks, 77–78
ROI. *See* Return on investment
Rummler/Brache Model, 85–86

Sales cycle, 153
Sales volume, 159–60
Sandia National Laboratories, 22–23
Saturn (company), 17
Schedule, 77
Self-assessment, 8
Self-managed teams, 9–10
Senge, Peter, 42, 47
Service
  basic, 65–66
  level of, 88
  training, 60–62
Service industries, 16
Seyfer, Charline, 23
Shultz, Jim, 160
Skill Dynamics, 19
Smile test, 146
Smith, Don, 59
Social pressures, 12–18
Social settings, 46
Staff
  development, 113–14
  teamwork, 14–15
Standards, 119–43
  available, 129–39
  criteria for, 121
  evaluation of, 125
  evolution of, 121–23
  IBSTPI, 25–26, 36, 50, 53, 56, 94, 123–24, 130–39
  inputs, 125–26
  instructional design, 138–39
  ISO, 4–12, 87, 124, 129–32, 136, 165
  models and examples for, 121
  outcomes, 128
  outputs, 127–28
  processes, 126–27, 135
  products, 126–27
  purpose of, 120–23
  U.K., 129–30, 133
  uses of, 123–28
Stewardship, 48–49

Strategy, 53–81
  Amoco ITT direction, 69–71
  business plan, 73–79
  challenge, 54–56
  competition, 55–60
  customer's needs, 60–62
  developing plan, 62–79
  mission statement, 62–63
  selecting, 67–73
  vision statement, 64–67
Structure, 71
Summative evaluation, 107, 148
Summative feedback, 86
Support, 66

Task analysis, 24, 89–91
Teamwork, 9–10, 14–15
Threats, 68
Trade associations, 19
Training
  and development, 58
  and leadership, 31–52
  manager, 138–39
  measurement and evaluation, 145–70
  optimizing processes for, 83–117
  organizational benefits that accrue from, 147, 153
  quality, 1–29
  services, 60–62
  standards, 119–43
  strategic planning, 53–81
Transactions, 159
Tuition reimbursement, 160–63
Turnover, 160–63

United Kingdom, 129–30, 133
Universities, 19

Values, 33–36
Vendor competence, 10–11
Video production, 21
Vision statement, 64–67

Walgreen Company, 39, 57–59, 65, 159–64
Wheatley, Margaret, 87
Whirlpool (company), 19
Women, 13
Work force, 12–13, 15–16